STUDENT SOLUTIONS MANUAL

CHAPTERS 20–42

PHYSICS

FOR SCIENTISTS AND ENGINEERS THIRD EDITION

A STRATEGIC APPROACH

Randall D. Knight

Larry Smith
Snow College

Brett Kraabel
PhD-Physics, University of Santa Barbara

PEARSON

Boston Columbus Indianapolis New York San Francisco Upper Saddle River
Amsterdam Cape Town Dubai London Madrid Milan Munich Paris Montréal Toronto
Delhi Mexico City São Paulo Sydney Hong Kong Seoul Singapore Taipei Tokyo

Publisher: James Smith
Senior Development Editor: Alice Houston, Ph.D.
Senior Project Editor: Martha Steele
Assistant Editor: Peter Alston
Media Producer: Kelly Reed
Senior Administrative Assistant: Cathy Glenn
Director of Marketing: Christy Lesko
Executive Marketing Manager: Kerry McGinnis
Managing Editor: Corinne Benson
Production Project Manager: Beth Collins
Manufacturing Buyer: Jeffrey Sargent
Production Management, Illustration, and Composition: PreMediaGlobal, Inc.
Cover Design: Riezebos-Holzbaur Design Group and Seventeenth Street Studios
Cover Photo Illustration: Riezebos-Holzbaur Design Group
Printer: Edwards Brothers Malloy

ISBN 13: 978-0-321-77269-5
ISBN 10: 0-321-77269-5

5 6 7 8 9 10—V069—19 18 17 16 15
www.pearsonhighered.com

Contents

Preface

This *student Solutions Manual* is intended to provide you with examples of good problem-solving techniques and strategies. To achieve that, the solutions presented here attempt to:

- Follow, in detail, the problem-solving strategies presented in the text.
- Articulate the reasoning that must be done before computation.
- Illustrate how to use drawings effectively.
- Demonstrate how to utilize graphs, ratios, units, and the many other "tactics" that must be successfully mastered and marshaled if a problem-solving strategy is to be effective.
- Show examples of assessing the reasonableness of a solution.
- Comment on the significance of a solution or on its relationship to other problems.

We recommend you try to solve each problem on your own before you read the solution. Simply reading solutions, without first struggling with the issues, has limited educational value.

As you work through each solution, make sure you understand how and why each step is taken. See if you can understand which aspects of the problem made this solution strategy appropriate. You will be successful on exams not by memorizing solutions to particular problems but by coming to recognize which kinds of problem-solving strategies go with which types of problems.

We have made every effort to be accurate and correct in these solutions. However, if you do find errors or ambiguities, we would be very grateful to hear from you. Please have your instructor contact the Pearson Education sales representative.

Acknowledgments for the First Edition

We are grateful for many helpful comments from Susan Cable, Randall Knight, and Steve Stonebraker. We express appreciation to Susan Emerson, who typed the word-processing manuscript, for her diligence in interpreting our handwritten copy. Finally, we would like to acknowledge the support from the Addison Wesley staff in getting the work into a publishable state. Our special thanks to Liana Allday, Alice Houston, and Sue Kimber for their willingness and preparedness in providing needed help at all times.

Pawan Kahol
Missouri State University

Donald Foster
Wichita State University

Acknowledgments for the Second Edition

I would like to acknowledge the patient support of my wife, Holly, who knows what is important.

Larry Smith
Snow College

I would like to acknowledge the assistance and support of my wife, Alice Nutter, who helped type many problems and was patient while I worked weekends.

Scott Nutter
Northern Kentucky University

Acknowledgments for the Third Edition

To Holly, Ryan, Timothy, Nathan, Tessa, and Tyler, who make it all worthwhile.

Larry Smith
Snow College

I gratefully acknowledge the assistance of the staff at Physical Sciences Communication.

Brett Kraabel
PhD-University of Santa Barbara

TRAVELING WAVES

Exercises and Problems

Section 20.1 The Wave Model

20.1. Model: The wave is a traveling wave on a stretched string.
Solve: The wave speed on a stretched string with linear density μ is

$$v_{\text{string}} = \sqrt{\frac{T_S}{\mu}} \Rightarrow 150 \text{ m/s} = \sqrt{\frac{75 \text{ N}}{\mu}} \Rightarrow \mu = 3.333 \times 10^{-3} \text{ kg/m}$$

For a wave speed of 180 m/s, the required tension will be

$$T_S = \mu v_{\text{string}}^2 = (3.333 \times 10^{-3} \text{ kg/m})(180 \text{ m/s})^2 = 110 \text{ N}$$

Section 20.2 One-Dimensional Waves

20.5. Model: This is a wave traveling at constant speed. The pulse moves 1 m to the left every second.
Visualize: The snapshot graph shows the wave at all points on the x-axis at $t = 2$ s. The leading edge of the wave—moving left—will reach $x = 0$ m 1 s later, at $t = 3$ s. The first part of the wave causes a sudden upward displacement at $t = 3$ s. The falling slope of the wave is 4 m wide, so it will take 4 s for the the displacement at $x = 0$ m to decrease from $+1$ cm to -1 cm. The trailing edge of the pulse arrives at $x = 0$ m at $t = 7$ s, which is 5 s after the figure given in the problem. The displacement now becomes zero and stays zero for all later times.

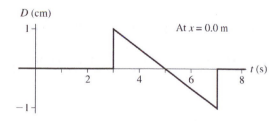

20.9. Visualize:

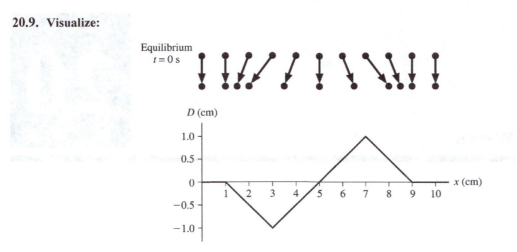

We first draw the particles of the medium in the equilibrium positions, with an inter-particle spacing of 1.0 cm. Just underneath, the positions of the particles as a longitudinal wave is passing through are shown at time $t = 0$ s. It is clear that relative to the equilibrium the particle positions are displaced negatively on the left side and positively on the right side. For example, the particles at $x = 0$ cm and $x = 1$ cm are at equilibrium, the particle at $x = 2$ cm is displaced left by 0.5 cm, the particle at $x = 3$ cm is displaced left by 1.0 cm, the particle at $x = 4$ cm is displaced left by 0.5 cm, and the particle at $x = 5$ cm is undisplaced. The behavior of particles for $x > 5$ cm is opposite of that for $x < 5$ cm.

Section 20.3 Sinusoidal Waves

20.11. Solve: (a) The wave number is

$$k = \frac{2\pi}{\lambda} = \frac{2\pi}{2.0 \text{ m}} = 3.1 \text{ rad/m}$$

(b) The wave speed is

$$v = \lambda f = \lambda \left(\frac{\omega}{2\pi} \right) = (2.0 \text{ m}) \left(\frac{30 \text{ rad/s}}{2\pi} \right) = 9.5 \text{ m/s}$$

Section 20.4 Waves in Two and Three Dimensions

20.15. Solve: According to Equation 20.28, the phase difference between two points on a wave is

$$\Delta \phi = \phi_2 - \phi_1 = \frac{2\pi}{\lambda}(r_2 - r_1)$$

If $\phi_1 = \pi$ rad at $r_1 = 4.0$ m, we can determine ϕ_2 at any r value at the same instant using this equation. At $r_2 = 3.5$ m,

$$\phi_2 = \phi_1 + \frac{2\pi}{\lambda}(r_2 - r_1) = \pi \text{ rad} + \frac{2\pi}{2.0 \text{ m}}(3.5 \text{ m} - 4.0 \text{ m}) = \frac{\pi}{2} \text{ rad}$$

At $r_2 = 4.5$ m, $\phi = \frac{3}{2}\pi$ rad.

20.17. Visualize: A phase difference of 2π rad corresponds to a distance of λ. Set up a ratio.
Solve:

$$\frac{x}{\lambda} = \frac{5.5 \text{ rad}}{2\pi \text{ rad}} \Rightarrow x = \left(\frac{5.5 \text{ rad}}{2\pi \text{ rad}} \right) \lambda = \left(\frac{5.5 \text{ rad}}{2\pi \text{ rad}} \right) \frac{v}{f} = \left(\frac{5.5 \text{ rad}}{2\pi \text{ rad}} \right) \frac{340 \text{ m/s}}{120 \text{ Hz}} = 2.5 \text{ m}$$

Assess: 2.5 m seems like a reasonable distance.

Section 20.5 Sound and Light

20.19. Solve: Two pulses of sound are detected because one pulse travels through the metal to the microphone while the other travels through the air to the microphone. The time interval for the sound pulse traveling through the air is

$$\Delta t_{air} = \frac{\Delta x}{v_{air}} = \frac{4.00 \text{ m}}{343 \text{ m/s}} = 0.01166 \text{ s} = 11.66 \text{ ms}$$

Sound travels *faster* through solids than gases, so the pulse traveling through the metal will reach the microphone *before* the pulse traveling through the air. Because the pulses are separated in time by 9.00 ms, the pulse traveling through the metal takes $\Delta t_{metal} = 2.66 \text{ ms}$ to travel the 4.00 m to the microphone. Thus, the speed of sound in the metal is

$$v_{metal} = \frac{\Delta x}{\Delta t_{metal}} = \frac{4.00 \text{ m}}{0.00266 \text{ s}} = 1504 \text{ m/s} \approx 1500 \text{ m/s}$$

20.21. Solve: (a) The frequency is

$$f = \frac{c}{\lambda} = \frac{3.0 \times 10^8 \text{ m/s}}{0.20 \text{ m}} = 1.5 \times 10^9 \text{ Hz} = 1.5 \text{ GHz}$$

(b) The speed of a sound wave in water is $v_{water} = 1480 \text{ m/s}$. The wavelength of the sound wave would be

$$\lambda = \frac{v_{water}}{f} = \frac{1480 \text{ m/s}}{1.50 \times 10^9 \text{ Hz}} = 9.87 \times 10^{-7} \text{ m} \approx 990 \text{ nm}$$

20.25. Model: Light is an electromagnetic wave.
Solve: (a) The time light takes is

$$t = \frac{3.0 \text{ mm}}{v_{glass}} = \frac{3.0 \times 10^{-3} \text{ m}}{c/n} = \frac{3.0 \times 10^{-3} \text{ m}}{(3.0 \times 10^8 \text{ m/s})/1.50} = 1.5 \times 10^{-11} \text{ s}$$

(b) The thickness of water is

$$d = v_{water}t = \frac{c}{n_{water}}t = \frac{3.0 \times 10^8 \text{ m/s}}{1.33}(1.5 \times 10^{-11} \text{ s}) = 3.4 \text{ mm}$$

Section 20.6 Power, Intensity, and Decibels

20.29. Solve: The energy delivered to an area a in time t is $E = Pt$, where the power P is related to the intensity I as $I = P/a$. Thus, the energy received by your back is

$$E = Pt = Iat = (0.80)(1400 \text{ W/m}^2)(0.30 \times 50 \text{ m}^2)(3600 \text{ s}) = 6.0 \times 10^5 \text{ J}$$

20.33. Visualize: Equation 20.35 gives the sound intensity level as

$$\beta = (10 \text{ dB})\log_{10}\left(\frac{I}{I_0}\right)$$

where $I_0 = 1.0 \times 10^{-12} \text{ W/m}^2$.
Solve:
(a)

$$\beta = (10 \text{ dB})\log_{10}\left(\frac{I}{I_0}\right) = (10 \text{ dB})\log_{10}\left(\frac{3.0 \times 10^{-6} \text{ W/m}^2}{1.0 \times 10^{-12} \text{ W/m}^2}\right) = 65 \text{ dB}$$

(b)

$$\beta = (10 \text{ dB})\log_{10}\left(\frac{I}{I_0}\right) = (10 \text{ dB})\log_{10}\left(\frac{3.0 \times 10^{-2} \text{ W/m}^2}{1.0 \times 10^{-12} \text{ W/m}^2}\right) = 105 \text{ dB}$$

Assess: As mentioned in the chapter, each factor of 10 in intensity changes the sound intensity level by 10 dB; between the first and second parts of this problem the intensity changed by a factor of 10^4, so we expect the sound intensity level to change by 40 dB.

20.35. Model: Assume the pole is tall enough that we don't have to worry about the ground absorbing or reflecting sound.
Visualize: The area of a sphere of radius R is $A = 4\pi R^2$. Also, $I = P/A$. We seek P when $R = 20$ m.
Solve:

$$P = IA = (I_0 \times 10^{\beta/10 \text{ dB}})(4\pi R^2) = (I_0 \times 10^{90 \text{ dB}/10 \text{ dB}})(4\pi (20 \text{ m})^2) = 5.0 \text{ W}$$

Assess: 5.0 W is a reasonable power output for a speaker.

Section 20.7 The Doppler Effect

20.37. Model: The frequency of the opera singer's note is altered by the Doppler effect.
Solve: (a) Using 90 km/h = 25 m/s, the frequency as her convertible approaches the stationary person is

$$f_+ = \frac{f_0}{1 - v_S/v} = \frac{600 \text{ Hz}}{1 - \dfrac{25 \text{ m/s}}{343 \text{ m/s}}} = 650 \text{ Hz}$$

(b) The frequency as her convertible recedes from the stationary person is

$$f_- = \frac{f_0}{1 + v_S/v} = \frac{600 \text{ Hz}}{1 + \dfrac{25 \text{ m/s}}{343 \text{ m/s}}} = 560 \text{ Hz}$$

20.41. Solve: (a) We see from the history graph that the period $T = 0.20$ s and the wave speed $v = 4.0$ m/s. Thus, the wavelength is

$$\lambda = \frac{v}{f} = vT = (4.0 \text{ m/s})(0.20 \text{ s}) = 0.80 \text{ m}$$

(b) The phase constant ϕ_0 is obtained as follows:

$$D(0 \text{ m}, 0 \text{ s}) = A\sin\phi_0 \Rightarrow 2 \text{ mm} = (2 \text{ mm})\sin\phi_0 \Rightarrow \sin\phi_0 = 1 \Rightarrow \phi_0 = \tfrac{1}{2}\pi \text{ rad}$$

(c) The displacement equation for the wave is

$$D(x, t) = A\sin\left(\frac{2\pi x}{\lambda} - 2\pi ft + \phi_0\right) = (2.0 \text{ mm})\sin\left(\frac{2\pi x}{0.80 \text{ m}} - \frac{2\pi t}{0.20 \text{ s}} + \frac{\pi}{2}\right) = (2.0 \text{ mm})\sin(2.5\pi x - 10\pi t + \tfrac{1}{2}\pi)$$

where x and t are in m and s, respectively.

20.45. Solve: Δt is the time the sound wave takes to travel down to the bottom of the ocean and then up to the ocean surface. The depth of the ocean is

$$2d = (v_{\text{sound in water}})\Delta t \Rightarrow d = (750 \text{ m/s})\Delta t$$

Using this relation and the data from Figure P20.45, we can generate the following table for the ocean depth (d) at various positions (x) of the ship.

x (km)	Δt (s)	d (km)
0	6	4.5
20	4	3.0
40	4	3.0
45	8	6.0
50	4	3.0
60	2	1.5

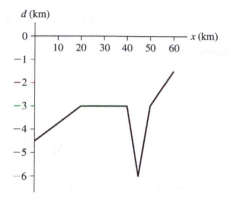

20.47. Model: Assume a room temperature of 20°C.
Visualize:

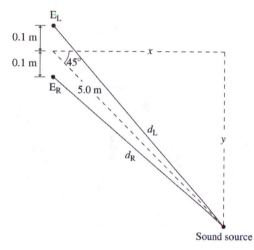

Sound source

Solve: The distance between the source and the left ear (E_L) is

$$d_L = \sqrt{x^2 + (y + 0.1 \text{ m})^2} = \sqrt{[(5.0 \text{ m})\cos 45°]^2 + [(5.0 \text{ m})\sin 45° + 0.1 \text{ m}]^2} = 5.0712 \text{ m}$$

Similarly $d_R = 4.9298$ m. Thus,

$$d_L - d_R = \Delta d = 0.1414 \text{ m}$$

For the sound wave with a speed of 343 m/s, the difference in arrival times at your left and right ears is

$$\Delta t = \frac{\Delta d}{343 \text{ m/s}} = \frac{0.1414 \text{ m}}{343 \text{ m/s}} = 410 \text{ } \mu s$$

20.51. Model: This is a sinusoidal wave.
Solve: (a) The equation is of the form $D(y, t) = A\sin(ky + \omega t + \phi_0)$, so the wave is traveling along the y-axis.
Because it is $+\omega t$ rather than $-\omega t$ the wave is traveling in the *negative* y-direction.
(b) Sound is a longitudinal wave, meaning that the medium is displaced *parallel* to the direction of travel. So the air
molecules are oscillating back and forth along the y-axis.
(c) The wave number is $k = 8.96$ m^{-1}, so the wavelength is

$$\lambda = \frac{2\pi}{k} = \frac{2\pi}{8.96 \text{ m}^{-1}} = 0.701 \text{ m}$$

The angular frequency is $\omega = 3140$ s^{-1}, so the wave's frequency is

$$f = \frac{\omega}{2\pi} = \frac{3140 \text{ s}^{-1}}{2\pi} = 500 \text{ Hz}$$

Thus, the wave speed $v = \lambda f = (0.70 \text{ m})(500 \text{ Hz}) = 350 \text{ m/s}$. The period $T = 1/f = 0.00200 \text{ s} = 2.00 \text{ ms}$.
Assess: The wave is a sound wave with speed $v = 350$ m/s. This is greater than the room-temperature speed of 343 m/s, so the air temperature must be greater than 20°.

20.53. Model: This is a sinusoidal wave traveling on a stretched string in the $+x$ direction.
Solve: (a) From the displacement equation of the wave, $A = 2.0$ cm, $k = 12.57$ rad/m, and $\omega = 638$ rad/s. Using the equation for the wave speed in a stretched string,

$$v_{\text{string}} = \sqrt{\frac{T_S}{\mu}} \Rightarrow T_S = \mu v_{\text{string}}^2 = \mu\left(\frac{\omega}{k}\right)^2 = (5.00\times10^{-3} \text{ kg/m}^3)\left(\frac{638 \text{ rad/s}}{12.57 \text{ rad/m}}\right)^2 = 12.6 \text{ N}$$

(b) The maximum displacement is the amplitude $D_{\max}(x,t) = 2.00$ cm.
(c) From Equation 20.17,

$$v_{y\,\max} = \omega A = (638 \text{ rad/s})(2.0\times10^{-2} \text{ m}) = 12.8 \text{ m/s}$$

20.57. Model: We have a wave traveling to the right on a string.
Visualize:

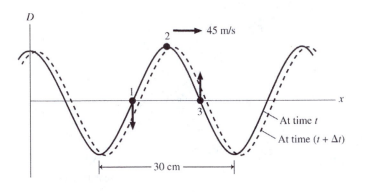

Solve: The snapshot of the wave as it travels to the right for an infinitesimally small time Δt shows that the velocity at point 1 is downward, at point 3 is upward, and at point 2 is zero. Furthermore, the speed at points 1 and 3 is the maximum speed given by Equation 20.17: $v_1 = v_3 = \omega A$. The frequency of the wave is

$$\omega = 2\pi f = 2\pi\frac{v}{\lambda} = \frac{2\pi(45 \text{ m/s})}{0.30 \text{ m}} = 300\pi \text{ rad/s} \Rightarrow \omega A = (300\pi \text{ rad/s})(2.0\times10^{-2} \text{ m}) = 19 \text{ m/s}$$

Thus, $v_1 = -19$ m/s, $v_2 = 0$ m/s, and $v_3 = +19$ m/s.

20.59. Solve: The wave speeds along the two metal wires are

$$v_1 = \sqrt{\frac{T}{\mu_1}} = \sqrt{\frac{2250 \text{ N}}{0.009 \text{ kg/m}}} = 500 \text{ m/s} \qquad v_2 = \sqrt{\frac{T}{\mu_2}} = \sqrt{\frac{2250 \text{ N}}{0.025 \text{ kg/m}}} = 300 \text{ m/s}$$

The wavelengths along the two wires are

$$\lambda_1 = \frac{v_1}{f} = \frac{500 \text{ m/s}}{1500 \text{ Hz}} = \frac{1}{3} \text{ m} \qquad \lambda_2 = \frac{v_2}{f} = \frac{300 \text{ m/s}}{1500 \text{ Hz}} = \frac{1}{5} \text{ m}$$

Thus, the number of wavelengths over two sections of the wire are

$$\frac{1.0 \text{ m}}{\lambda_1} = \frac{1.0 \text{ m}}{\left(\frac{1}{3} \text{ m}\right)} = 3 \qquad \frac{1.0 \text{ m}}{\lambda_2} = \frac{1.0 \text{ m}}{\left(\frac{1}{5} \text{ m}\right)} = 5$$

The number of complete cycles of the wave in the 2.00-m-long wire is 8.

20.61. Model: The wave is traveling on a stretched string.
Solve: The wave speed on the string is

$$v = \sqrt{\frac{T_S}{\mu}} = \sqrt{\frac{50 \text{ N}}{0.005 \text{ kg/m}}} = 100 \text{ m/s}$$

The speed of the particle on the string, however, is given by Equation 20.17. The maximum speed is calculated as follows:

$$v_y = -\omega A \cos(kx - \omega t + \phi_0) \Rightarrow v_{y \text{ max}} = \omega A = 2\pi f A = 2\pi \frac{v}{\lambda} A = 2\pi \left(\frac{100 \text{ m/s}}{2.0 \text{ m}}\right)(0.030 \text{ m}) = 9.4 \text{ m/s}$$

20.65. Visualize: To find the power of a laser pulse, we need the energy it contains, U, and the time duration of the pulse, Δt. Then to find the intensity, we need the area of the pulse. Its radius is 0.50 mm.
Solve: (a) Using $P = U/\Delta t$, we find the following:

$$P = (1.0 \times 10^{-3} \text{ J})/(15 \times 10^{-9} \text{ s}) = 6.67 \times 10^4 \text{ W}$$

(b) Then from $I = P/a$, we obtain

$$I = \frac{(6.67 \times 10^4 \text{ W})}{\pi (5.0 \times 10^{-4} \text{ m})^2} = 8.5 \times 10^{10} \text{ W/m}^2$$

20.67. Model: Assume the saw is far enough off the ground that we don't have to worry about reflected sound.
Visualize: First note that $\beta_1 - \beta_2 = 20 \text{ dB} \Rightarrow I_1/I_2 = 10 \cdot 10 = 100$ (a change of 10 dB corresponds to a change in intensity by a factor of 10). Then use $I_1 A_1 = P$ and then $P = I_2 A_2 \Rightarrow A_2 = P/I_2$, and finally solve for $R_2 = \sqrt{A_2/4\pi}$.
Solve: Put all of the above together.

$$R_2 = \sqrt{\frac{A_2}{4\pi}} = \sqrt{\frac{\frac{P}{I_2}}{4\pi}} = \sqrt{\frac{\frac{I_1 A_1}{I_2}}{4\pi}} = \sqrt{\frac{\frac{I_1}{I_2}(4\pi R_1^2)}{4\pi}} = R_1 \sqrt{\frac{I_1}{I_2}} = R_1 \sqrt{100} = (5.0 \text{ m})(10) = 50 \text{ m}$$

Assess: The scaling laws help and the answer is reasonable.

20.69. Solve: If we solve the equation for I, we have:

$$I = I_0 \times 10^{(\beta/10 \text{ dB})}$$

Now plugging in 60 dB for β, we get $I = 10^{-6} \text{ W/m}^2$ and plugging in 61 dB for β, we get $I = 1.3 \times 10^{-6} \text{ W/m}^2$. The ratio of the latter to the former is 1.3.

20.75. Model: The Doppler effect for light of a receding source yields an increased wavelength.
Solve: Because the measured wavelengths are 0.5% longer, that is, $\lambda = 1.005\lambda_0$, the distant galaxy is receding away from the earth. Using Equation 20.40,

$$\lambda = 1.005\lambda_0 = \sqrt{\frac{1 + v_s/c}{1 - v_s/c}}\lambda_0 \Rightarrow (1.005)^2 = \frac{1 + v_s/c}{1 - v_s/c} \Rightarrow v_s = 0.0050 \, c = 1.5 \times 10^6 \text{ m/s}$$

SUPERPOSITION

Exercises and Problems

Section 21.1 The Principle of Superposition

21.1. Model: The principle of superposition comes into play whenever the waves overlap.
Visualize:

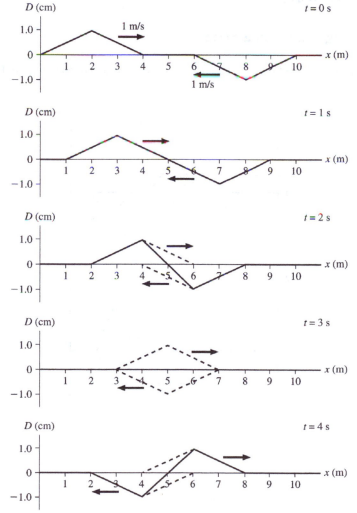

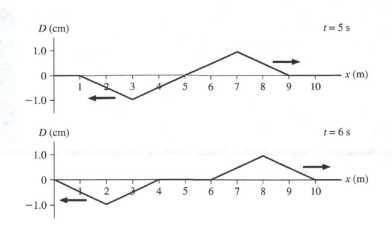

The snapshot graph at $t = 1.0$ s differs from the graph $t = 0.0$ s in that the left wave has moved to the right by 1.0 m and the right wave has moved to the left by 1.0 m. This is because the distance covered by each wave in 1.0 s is 1.0 m. The snapshot graphs at $t = 2.0$, 3.0, and 4.0 s are a superposition of the left and the right moving waves. The overlapping parts of the two waves are shown by the dotted lines.

Section 21.2 Standing Waves

Section 21.3 Standing Waves on a String

21.5. Model: A wave pulse reflected from the string-wall boundary is inverted and its amplitude is unchanged.
Visualize:

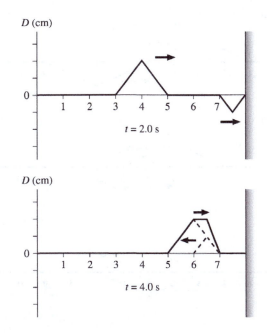

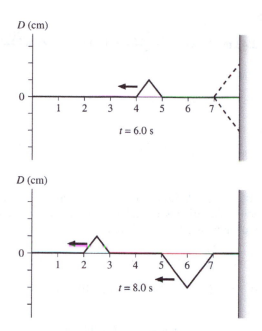

The graph at $t = 2$ s differs from the graph at $t = 0$ s in that both waves have moved to the right by 2 m. This is because the distance covered by the wave pulse in 2 s is 2 m. The shorter pulse wave encounters the boundary wall at 2 s and is inverted upon reflection. This reflected pulse wave overlaps with the broader pulse wave, as shown in the snapshot graph at $t = 4$ s. At $t = 6$ s, only half of the broad pulse is reflected and hence inverted; the shorter pulse wave continues to move to the left with a speed of 1 m/s. Finally, at $t = 8$ s both the reflected pulse waves are inverted and they are both moving to the left.

21.7. Model: Reflections at both ends of the string cause the formation of a standing wave.
Solve: Figure EX21.7 indicates 5/2 wavelengths on the 2.0-m-long string. Thus, the wavelength of the standing wave is $\lambda = \frac{2}{5}(2.0 \text{ m}) = 0.80$ m. The frequency of the standing wave is

$$f = \frac{v}{\lambda} = \frac{40 \text{ m/s}}{0.80 \text{ m}} = 50 \text{ Hz}$$

21.9. Model: A string fixed at both ends supports standing waves.
Solve: (a) A standing wave can exist on the string only if its wavelength is

$$\lambda_m = \frac{2L}{m} \qquad m = 1, 2, 3, \ldots$$

The three longest wavelengths for standing waves will therefore correspond to $m = 1, 2$, and 3. Thus,

$$\lambda_1 = \frac{2(2.4 \text{ m})}{1} = 4.8 \text{ m} \quad \lambda_2 = \frac{2(2.4 \text{ m})}{2} = 2.4 \text{ m} \quad \lambda_3 = \frac{2(2.4 \text{ m})}{3} = 1.6 \text{ m}$$

(b) Because the wave speed on the string is unchanged from one m value to the other,

$$f_2 \lambda_2 = f_3 \lambda_3 \quad \Rightarrow \quad f_3 = \frac{f_2 \lambda_2}{\lambda_3} = \frac{(50 \text{ Hz})(2.4 \text{ m})}{1.6 \text{ m}} = 75 \text{ Hz}$$

Section 21.4 Standing Sound Waves and Musical Acoustics

21.13. Solve: (a) For the open-open tube, the two open ends exhibit antinodes of a standing wave. The possible wavelengths for this case are

$$\lambda_m = \frac{2L}{m} \qquad m = 1, 2, 3, \ldots$$

The three longest wavelengths are

$$\lambda_1 = \frac{2(1.21\ m)}{1} = 2.42\ m \qquad \lambda_2 = \frac{2(1.21\ m)}{2} = 1.21\ m \qquad \lambda_3 = \frac{2(1.21\ m)}{3} = 0.807\ m$$

(b) In the case of an open-closed tube,

$$\lambda_m = \frac{4L}{m} \quad m = 1,\ 3,\ 5,\ \ldots$$

The three longest wavelengths are

$$\lambda_1 = \frac{4(1.21\ m)}{1} = 4.84\ m \qquad \lambda_2 = \frac{4(1.21\ m)}{3} = 1.61\ m \qquad \lambda_3 = \frac{4(1.21\ m)}{5} = 0.968\ m$$

21.17. Model: Reflections at the string boundaries cause a standing wave on a stretched string.
Solve: Because the vibrating section of the string is 1.9 m long, the two ends of this vibrating wire are fixed, and the string is vibrating in the fundamental harmonic. The wavelength is

$$\lambda_m = \frac{2L}{m} \quad \Rightarrow \quad \lambda_1 = 2L = 2(1.90\ m) = 3.80\ m$$

The wave speed along the string is $v = f_1\lambda_1 = (27.5\ Hz)(3.80\ m) = 104.5\ m/s$. The tension in the wire can be found as follows:

$$v = \sqrt{\frac{T_S}{\mu}} \quad \Rightarrow \quad T_S = \mu v^2 = \left(\frac{mass}{length}\right)v^2 = \left(\frac{0.400\ kg}{2.00\ m}\right)(104.5\ m/s)^2 = 2180\ N$$

Section 21.5 Interference in One Dimension

Section 21.6 The Mathematics of Interference

21.19. Model: Interference occurs according to the difference between the phases $(\Delta\phi)$ of the two waves.
Solve: (a) A separation of 20 cm between the speakers leads to maximum intensity on the x-axis, but a separation of 60 cm leads to zero intensity. That is, the waves are in phase when $(\Delta x)_1 = 20$ cm but out of phase when $(\Delta x)_2 = 60$ cm. Thus,

$$(\Delta x)_2 - (\Delta x)_1 = \frac{\lambda}{2} \quad \Rightarrow \quad \lambda = 2(60\ cm - 20\ cm) = 80\ cm$$

(b) If the distance between the speakers continues to increase, the intensity will again be a maximum when the separation between the speakers that produced a maximum has increased by one wavelength. That is, when the separation between the speakers is $20\ cm + 80\ cm = 100\ cm$.

21.21. Model: Reflection is maximized if the two reflected waves interfere constructively.
Solve: The film thickness that causes constructive interference at wavelength λ is given by Equation 21.32:

$$\lambda_C = \frac{2nd}{m} \quad \Rightarrow \quad d = \frac{\lambda_C m}{2n} = \frac{(600 \times 10^{-9}\ m)(1)}{(2)(1.39)} = 216\ nm$$

where we have used $m = 1$ to calculate the thinnest film.
Assess: The film thickness is much less than the wavelength of visible light. The above formula is applicable because $n_{air} < n_{film} < n_{glass}$.

Section 21.7 Interference in Two and Three Dimensions

21.25. Model: The two speakers are identical, and so they are emitting circular waves in phase. The overlap of these waves causes interference.

Visualize:

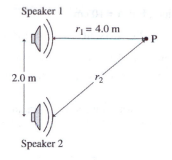

Speaker 1

$r_1 = 4.0$ m

P

2.0 m

r_2

Speaker 2

Solve: From the geometry of the figure,

$$r_2 = \sqrt{r_1^2 + (2.0 \text{ m})^2} = \sqrt{(4.0 \text{ m})^2 + (2.0 \text{ m})^2} = 4.472 \text{ m}$$

so $\Delta r = r_2 - r_1 = 4.472$ m $- 4.0$ m $= 0.472$ m. The phase difference between the sources is $\Delta\phi_0 = 0$ rad and the wavelength of the sound waves is

$$\lambda = \frac{v}{f} = \frac{340 \text{ m/s}}{1800 \text{ Hz}} = 0.1889 \text{ m}$$

Thus, the phase difference of the waves at the point 4.0 m in front of one source is

$$\Delta\phi = 2\pi \frac{\Delta r}{\lambda} + \Delta\phi_0 = \frac{2\pi(0.472 \text{ m})}{0.1889 \text{ m}} + 0 \text{ rad} = 5\pi \text{ rad} = \frac{5}{2}(2\pi \text{ rad})$$

This is a half-integer multiple of 2π rad, so the interference is perfect destructive.

Section 21.8 Beats

21.29. Solve: The beat frequency is

$$f_{\text{beat}} = f_1 - f_2 = 100 \text{ MHz}$$

where we have used the fact that the $\lambda_1 < \lambda_2$ so $f_1 > f_2$. The frequency of emitter 1 is $f_1 = c/\lambda_1$, where $\lambda_1 = 1.250 \times 10^{-2}$ m. The wavelength of emitter 2 is

$$\lambda_2 = c/f_2 = \frac{c}{f_1 - 100 \text{ MHz}} = \frac{c}{c/\lambda_1 - 100 \text{ MHz}} = \frac{(3.00 \times 10^8 \text{ m/s})}{(3.00 \times 10^8 \text{ m/s})/(1.250 \times 10^{-2} \text{ m}) - 100 \text{ MHz}} = 1.26 \text{ cm}.$$

21.31. Model: The wavelength of the standing wave on a string vibrating at its second-harmonic frequency is equal to the string's length.
Visualize:

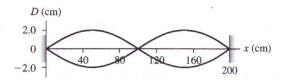

D (cm)

2.0

0

−2.0

40   80   120   160

200

x (cm)

Solve: The length of the string $L = 2.0$ m, so $\lambda = L = 2.0$ m. This means the wave number is

$$k = \frac{2\pi}{\lambda} = \frac{2\pi}{2.0 \text{ m}} = \pi \text{ rad/m}$$

According to Equation 21.5, the displacement of a medium when two sinusoidal waves superpose to give a standing wave is $D(x,t) = A(x)\cos\omega t$, where $A(x) = 2a\sin kx = A_{max}\sin kx$. The amplitude function gives the amplitude of oscillation from point to point in the medium. For $x = 10$ cm,

$$A(x = 10 \text{ cm}) = (2.0 \text{ cm})\sin[(\pi \text{ rad/m})(0.10 \text{ m})] = 0.62 \text{ cm}$$

Similarly, $A(x = 20 \text{ cm}) = 1.2$ cm, $A(x = 30 \text{ cm}) = 1.6$ cm, $A(x = 40 \text{ cm}) = 1.9$ cm, and $A(x = 50 \text{ cm}) = 2.0$ cm.

Assess: Consistent with the above figure, the amplitude of oscillation is a maximum at $x = 0.50$ m.

21.33. Model: The wavelength of the standing wave on a string vibrating at its fundamental frequency is equal to $2L$.
Solve: The amplitude of oscillation on the string is $A(x) = 2a\sin kx$, where a is the amplitude of the traveling wave and the wave number is

$$k = \frac{2\pi}{\lambda} = \frac{2\pi}{2L} = \frac{\pi}{L}$$

Substituting k into the above equation gives

$$A(x = \tfrac{1}{4}L) = 2.0 \text{ cm} = 2a\sin\left[\left(\frac{\pi}{L}\right)\left(\frac{L}{4}\right)\right] \implies 1.0 \text{ cm} = a\left(\frac{1}{\sqrt{2}}\right) \implies a = \sqrt{2} \text{ cm} = 1.4 \text{ cm}$$

21.37. Model: The wave on a stretched string with both ends fixed is a standing wave. For a vibration at its fundamental frequency, $\lambda = 2L$.
Solve: The wavelength of the wave reaching your ear is 39.1 cm $= 0.391$ m, so the frequency of the sound wave is

$$f = \frac{v_{air}}{\lambda} = \frac{344 \text{ m/s}}{0.391 \text{ m}} = 879.8 \text{ Hz}$$

This is also the frequency emitted by the wave on the string. Thus,

$$879.8 \text{ Hz} = \frac{v_{string}}{\lambda} = \frac{1}{\lambda}\sqrt{\frac{T_S}{\mu}} = \frac{1}{\lambda}\sqrt{\frac{150 \text{ N}}{0.0006 \text{ kg/m}}} \implies \lambda = 0.568 \text{ m}$$

$$L = \tfrac{1}{2}\lambda = 0.284 \text{ m} = 28.4 \text{ cm}$$

21.41. Model: The stretched bungee cord that forms a standing wave with two antinodes is vibrating at the second harmonic frequency.
Visualize:

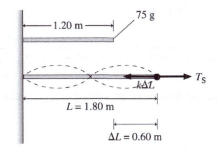

Solve: Because the vibrating cord has two antinodes, $\lambda_2 = L = 1.80$ m. The wave speed on the cord is

$$v_{cord} = f\lambda = (20 \text{ Hz})(1.80 \text{ m}) = 36 \text{ m/s}$$

The linear density of the cord is $v_{cord} = \sqrt{T_S/\mu}$. The tension T_S in the cord is equal to $k\Delta L$, where k is the bungee's spring constant and ΔL is the 0.60 m the bungee has been stretched. The linear density has to be calculated at the stretched length of 1.8 m where it is now vibrating. Thus,

$$v_{cord} = \sqrt{\frac{T_S}{\mu}} = \sqrt{\frac{k\Delta L}{m/L}} \implies k = \frac{mv_{cord}^2}{L\Delta L} = \frac{(0.075 \text{ kg})(36 \text{ m/s})^2}{(1.80 \text{ m})(0.60 \text{ m})} = 90 \text{ N/m}$$

21.43. Visualize: Use primed quantities for when the sphere is submerged. We are given $f_5' = f_3$ and $M = 1.5$ kg. We also know the density of water is $\rho = 1000$ kg/m^3. In the third mode before the sphere is submerged $L = \frac{3}{2}\lambda \Rightarrow \lambda = \frac{2}{3}L$. Likewise, after the sphere is submerged $L = \frac{5}{2}\lambda' \Rightarrow \lambda' = \frac{2}{5}L$. The tension in the string before the sphere is submerged is $T_s = Mg$, but after the sphere is submerged, according to Archimedes' principle, it is reduced by the weight of the water displaced by the sphere: $T_s' = Mg - \rho Vg$, where $V = \frac{4}{3}\pi R^3$.

Solve: We are looking for R so solve $T_s' = Mg - \rho Vg$ for ρVg and later we will isolate R from that.

$$\rho Vg = Mg - T_0'$$

Solve $v' = \sqrt{T_s/\mu}$ for T_s'. Also substitute for V.

$$\rho\left(\frac{4}{3}\right)\pi R^3 g = Mg - \mu v'^2$$

Now use $v' = \lambda' f'$.

$$\rho\left(\frac{4}{3}\right)\pi R^3 g = Mg - \mu(\lambda' f_5')^2$$

Recall that $f_5' = f_3$ and $\lambda' = \frac{2}{5}L$.

$$\rho\left(\frac{4}{3}\right)\pi R^3 g = Mg - \mu\left(\frac{2}{5}Lf_3\right)^2$$

Substitute $f_3 = v/\lambda$.

$$\rho\left(\frac{4}{3}\right)\pi R^3 g = Mg - \mu\left(\frac{2}{5}L\frac{v}{\lambda}\right)^2$$

Now use $\lambda = \frac{2}{3}L$ and $v = \sqrt{T_s/\mu}$.

$$\rho\left(\frac{4}{3}\right)\pi R^3 g = Mg - \mu\left(\frac{2}{5}L\frac{\sqrt{T_s/\mu}}{\frac{2}{3}L}\right)^2$$

The 2's, μ's, and L's cancel.

$$\rho\left(\frac{4}{3}\right)\pi R^3 g = Mg - \left(\frac{3}{5}\sqrt{T_s}\right)^2$$

Recall that $T_s = Mg$.

$$\rho\left(\frac{4}{3}\right)\pi R^3 g = Mg - \left(\frac{3}{5}\right)^2 Mg$$

Cancel g and factor M out on the right side.

$$\rho\left(\frac{4}{3}\right)\pi R^3 = M\left[1 - \left(\frac{3}{5}\right)^2\right] = M\left(1 - \frac{9}{25}\right) = M\left(\frac{16}{25}\right)$$

Now solve for R.

$$R^3 = \left(\frac{3}{4}\right)\left(\frac{16}{25}\right)\frac{M}{\pi\rho}$$

$$R = \sqrt[3]{\frac{12}{25}\frac{M}{\pi\rho}} = \sqrt[3]{\frac{12}{25}\frac{(1.5 \text{ kg})}{\pi(1000 \text{ kg/m}^3)}} = 6.1 \text{ cm}$$

Assess: The density of the sphere turns out to be about 1.5 times the density of water, which means it sinks and is in a reasonable range for densities.

21.47. Model: The fundamental wavelength of an open-open tube is $2L$ and that of an open-closed tube is $4L$.
Solve: We are given that

$$f_{1\,\text{open-closed}} = f_{3\,\text{open-open}} = 3f_{1\,\text{open-open}}$$

$$\frac{v_{\text{air}}}{\lambda_{1\,\text{open-closed}}} = 3\frac{v_{\text{air}}}{\lambda_{1\,\text{open-open}}} \quad \Rightarrow \quad \frac{1}{4L_{\text{open-closed}}} = \frac{3}{2L_{\text{open-open}}}$$

$$L_{\text{open-closed}} = \frac{2L_{\text{open-open}}}{12} = \frac{2(78.0\ \text{cm})}{12} = 13.0\ \text{cm}$$

21.51. Model: The nodes of a standing wave are spaced $\lambda/2$ apart. Assume there are no standing-wave modes between those given.
Visualize:

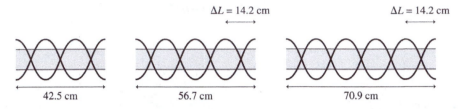

$$\Delta L = 14.2\ \text{cm} \qquad\qquad \Delta L = 14.2\ \text{cm}$$

42.5 cm   56.7 cm   70.9 cm

Solve: The wavelength of the mth mode of an open-open tube is $\lambda_m = 2L/m$. Or, equivalently, the length of the tube that generates the mth mode is $L = m(\lambda/2)$. Here λ is the same for all modes because the frequency of the tuning fork is unchanged. Increasing the length of the tube to go from mode m to mode $m+1$ requires a length change

$$\Delta L = (m+1)(\lambda/2) - m\lambda/2 = \lambda/2$$

That is, lengthening the tube by $\lambda/2$ adds an additional antinode and creates the next standing wave. This is consistent with the idea that the nodes of a standing wave are spaced $\lambda/2$ apart. This tube is first increased $\Delta L = 56.7\ \text{cm} - 42.5\ \text{cm} = 14.2\ \text{cm}$, then by $\Delta L = 70.9\ \text{cm} - 56.7\ \text{cm} = 14.2\ \text{cm}$. Thus $\lambda/2 = 14.2\ \text{cm}$ and thus $\lambda = 28.4\ \text{cm} = 0.284\ \text{m}$. Therefore the frequency of the tuning fork is

$$f = \frac{v}{\lambda} = \frac{343\ \text{m/s}}{0.284\ \text{m}} = 1208\ \text{Hz} \approx 12.1\ \text{kHz}$$

21.53. Model: A stretched wire, which is fixed at both ends, creates a standing wave whose fundamental frequency is $f_{1\,\text{wire}}$. The second vibrational mode of an open-closed tube is $f_{3\,\text{open-closed}}$. These two frequencies are equal because the wire's vibrations generate the sound wave in the open-closed tube.
Visualize:

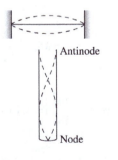

Antinode

Node

Solve: The frequency in the tube is

$$f_{3\,\text{open-closed}} = \frac{3v_{\text{air}}}{4L_{\text{tube}}} = \frac{3(340\ \text{m/s})}{4(0.85\ \text{cm})} = 300\ \text{Hz}$$

$$f_{1\,\text{wire}} = 300\ \text{Hz} = \frac{v_{\text{wire}}}{2L_{\text{wire}}} = \frac{1}{2L_{\text{wire}}}\sqrt{\frac{T_S}{\mu}}$$

$$T_S = (300\ \text{Hz})^2(2L_{\text{wire}})^2\mu = (300\ \text{Hz})^2(2\times0.25\ \text{m})^2(0.020\ \text{kg/m}) = 450\ \text{N}$$

21.55. Model: Model the tunnel as an open-closed tube.

Visualize: We are given $v = 335$ m/s. We would like to use $f_m = m\dfrac{v}{4L}$ $(m = \text{odd})$ to find L, but we need to know m first. Since m takes on only odd values for the open-closed tube the next resonance after m is $m + 2$. We are given $f_m = 4.5$ Hz and $f_{m+2} = 6.3$ Hz.

Solve:

$$\frac{f_{m+2}}{f_m} = \frac{(m+2)\dfrac{v}{4L}}{(m)\dfrac{v}{4L}} = \frac{m+2}{m} \;\Rightarrow\; m\!\left(\frac{f_{m+2}}{f_m}\right) = m+2 \;\Rightarrow\; m\!\left(\frac{f_{m+2}}{f_m} - 1\right) = 2 \;\Rightarrow\; m = \frac{2}{\dfrac{f_{m+2}}{f_m} - 1} = \frac{2}{\dfrac{6.3\ \text{Hz}}{4.5\ \text{Hz}} - 1} = 5$$

Now that we know m we can find the length L of the tunnel.

$$f_m = m\frac{v}{4L} \;\Rightarrow\; L = m\frac{v}{4f_m} = (5)\frac{335\ \text{m/s}}{4(4.5\ \text{Hz})} = 93\ \text{m}$$

Assess: 93 m seems like a reasonable length for a tunnel.

21.59. Model: Interference occurs according to the difference between the phases of the two waves.
Visualize:

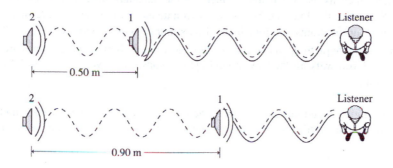

Solve: (a) The phase difference between the sound waves from the two speakers is

$$\Delta\phi = 2\pi\frac{\Delta x}{\lambda} + \Delta\phi_0$$

We have a maximum intensity when $\Delta x = 0.50$ m and $\Delta x = 0.90$ m. This means

$$2\pi\frac{(0.50\ \text{m})}{\lambda} + \Delta\phi_0 = 2m\pi\ \text{rad} \qquad 2\pi\!\left(\frac{0.90\ \text{m}}{\lambda}\right) + \Delta\phi_0 = 2(m+1)\pi\ \text{rad}$$

Taking the difference of these two equations gives

$$2\pi\!\left(\frac{0.40\ \text{m}}{\lambda}\right) = 2\pi \;\Rightarrow\; \lambda = 0.40\ \text{m} \;\Rightarrow\; f = \frac{v_{\text{sound}}}{\lambda} = \frac{340\ \text{m/s}}{0.40\ \text{m}} = 850\ \text{Hz}$$

(b) Using again the equations that correspond to constructive interference, we find

$$2\pi\!\left(\frac{0.50\ \text{m}}{0.40\ \text{m}}\right) + \Delta\phi_0 = 2m\pi\ \text{rad} \;\Rightarrow\; \Delta\phi_0 = \phi_{20} - \phi_{10} = -\frac{\pi}{2}\,\text{rad}$$

We have taken $m = 1$ in the last equation. This is because we always specify phase constants in the range $-\pi$ rad to π rad (or 0 rad to 2π rad). $m = 1$ gives $-\frac{1}{2}\pi$ rad (or equivalently, $m = 2$ will give $\frac{3}{2}\pi$ rad).

21.63. Model: The two radio antennas are sources of in-phase, circular waves. The overlap of these waves causes interference.

Visualize:

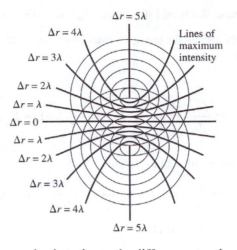

Solve: Maxima occur along lines such that the path difference to the two antennas is $\Delta r = m\lambda$. The 750 MHz $= 7.50 \times 10^8$ Hz wave has a wavelength $\lambda = c/f = 0.40$ m. Thus, the antenna spacing $d = 2.0$ m is exactly 5λ. The maximum possible intensity is on the line connecting the antennas, where $\Delta r = d = 5\lambda$. So this is a line of maximum intensity. Similarly, the line that bisects the two antennas is the $\Delta r = 0$ line of maximum intensity. In between, in each of the four quadrants, are four lines of maximum intensity with $\Delta r = \lambda, 2\lambda, 3\lambda,$ and 4λ. Although we have drawn a fairly accurate picture, you do *not* need to know precisely where these lines are located to know that you *have* to cross them if you walk all the way around the antennas. Thus, you will cross 20 lines where $\Delta r = m\lambda$ and will detect 20 maxima.

21.65. Model: The changing sound intensity is due to the interference of two overlapped sound waves.
Visualize: The listener moving relative to the speakers changes the phase difference between the waves.

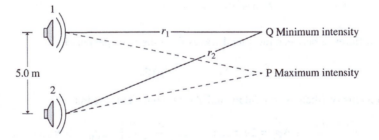

Solve: Initially when you are at P, equidistant from the speakers, you hear a sound of maximum intensity. This implies that the two speakers are in phase ($\Delta\phi_0 = 0$). However, upon moving to Q you hear a minimum of sound intensity, which implies that the path length difference from the two speakers to Q is $\lambda/2$. Thus,

$$\tfrac{1}{2}\lambda = \Delta r = r_2 - r_1 = \sqrt{(r_1)^2 + (5.0 \text{ m})^2} - r_1 = \sqrt{(12.0 \text{ m})^2 + (5.0 \text{ m})^2} - 12.0 \text{ m} = 1.0 \text{ m}$$

$$\lambda = 2.0 \text{ m} \quad \Rightarrow \quad f = \frac{v}{\lambda} = \frac{340 \text{ m/s}}{2.0 \text{ m}} = 170 \text{ Hz}$$

21.67. Model: The amplitude is determined by the interference of the two waves.
Visualize:

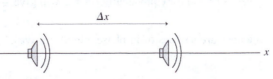

Solve: The amplitude of the sound wave is $A = \left| 2a\cos\left(\frac{1}{2}\Delta\phi\right) \right|$. We are to find the minimum ratio $\Delta x/\lambda$ at which $A = a$. Because the speakers are identical, we may take $\Delta\phi_0 = 0$.

$$a = 2a\cos\left(\frac{\Delta\phi}{2}\right) \quad \Rightarrow \quad \Delta\phi = 2\cos^{-1}\left(\tfrac{1}{2}\right) = 2\pi\frac{\Delta x}{\lambda} \quad \Rightarrow \quad \frac{\Delta x}{\lambda} = \frac{1}{\pi}\cos^{-1}\left(\tfrac{1}{2}\right) = \frac{1}{3}$$

21.71. Model: The superposition of two slightly different frequencies creates beats.
Solve: (a) The wavelength of the sound initially created by the flutist is

$$\lambda = \frac{342 \text{ m/s}}{440 \text{ Hz}} = 0.77727 \text{ m}$$

When the speed of sound inside her flute has increased due to the warming up of the air, the new frequency of the A note is

$$f' = \frac{346 \text{ m/s}}{0.77727 \text{ m}} = 445 \text{ Hz}$$

Thus the flutist will hear beats at the following frequency:

$$f' - f = 445 \text{ Hz} - 440 \text{ Hz} = 5 \text{ beats/s}$$

Note that the wavelength of the A note is determined by the length of the flute rather than the temperature of air or the increased sound speed.
(b) The initial length of the flute is $L = \frac{1}{2}\lambda = \frac{1}{2}(0.77727 \text{ m}) = 0.3886 \text{ m}$. The new length to eliminate beats needs to be

$$L' = \frac{\lambda'}{2} = \frac{1}{2}\left(\frac{v'}{f}\right) = \frac{1}{2}\left(\frac{346 \text{ m/s}}{440 \text{ Hz}}\right) = 0.3932 \text{ m}$$

Thus, she will have to extend the "tuning joint" of her flute by

$$0.3932 \text{ m} - 0.3886 \text{ m} = 0.0046 \text{ m} = 4.6 \text{ mm}$$

21.73. Model: The frequency of the loudspeaker's sound in the back of the pick-up truck is Doppler shifted. As the truck moves away from you, the frequency of the sound emitted by its speaker is decreased.
Solve: Because you hear 8 beats per second as the truck drives away from you, the frequency of the sound from the speaker in the pick-up truck is $f_- = 400 \text{ Hz} - 8 \text{ Hz} = 392 \text{ Hz}$. This frequency is

$$f_- = \frac{f_0}{1 + v_S/v} \quad \Rightarrow \quad 1 + \frac{v_S}{343 \text{ m/s}} = \frac{400 \text{ Hz}}{392 \text{ Hz}} = 1.020408 \quad \Rightarrow \quad v_S = 7.0 \text{ m/s}$$

That is, the velocity of the source v_S and hence the pick-up truck is 7.0 m/s.

WAVE OPTICS

Exercises and Problems

Section 22.2 The Interference of Light

22.1. Model: Two closely spaced slits produce a double-slit interference pattern.
Visualize: The interference pattern looks like the photograph of Figure 22.3(b). It is symmetrical with the $m = 3$ fringes on both sides of and equally distant from the central maximum.
Solve: The bright fringes occur at angles θ_m such that

$$d\sin\theta_m = m\lambda \quad m = 0, 1, 2, 3, \ldots$$

$$\Rightarrow \sin\theta_3 = \frac{3(600\times10^{-9}\ \text{m})}{(80\times10^{-6}\ \text{m})} = 0.0225 \Rightarrow \theta_3 = 0.023\ \text{rad} = 0.023\ \text{rad}\times\frac{180°}{\pi\ \text{rad}} = 1.3°$$

22.5. Visualize: The fringe spacing for a double slit pattern is $\Delta y = \dfrac{\lambda L}{d}$. We are given $L = 2.0$ m and $\lambda = 600$ nm.
We also see from the figure that $\Delta y = \frac{1}{3}$cm.
Solve: Solve the equation for d.

$$d = \frac{\lambda L}{\Delta y} = \frac{(600\times10^{-9}\ \text{m})(2.0\ \text{m})}{\frac{1}{3}\times10^{-2}\ \text{m}} = 0.36\ \text{mm}$$

Assess: 0.36 mm is a typical slit spacing.

22.7. Model: Two closely spaced slits produce a double-slit interference pattern.
Visualize: The interference fringes are equally spaced on both sides of the central maximum. The interference pattern looks like Figure 22.3(b).
Solve: In the small-angle approximation

$$\Delta\theta = \theta_{m+1} - \theta_m = (m+1)\frac{\lambda}{d} - m\frac{\lambda}{d} = \frac{\lambda}{d}$$

Since $d = 200\lambda$, we have

$$\Delta\theta = \frac{\lambda}{d} = \frac{1}{200}\ \text{rad} = 0.286°$$

Section 22.3 The Diffraction Grating

22.11. Model: A diffraction grating produces an interference pattern.
Visualize: The interference pattern looks like the diagram in Figure 22.8.

Solve: The bright constructive-interference fringes are given by Equation 22.15:

$$d\sin\theta_m = m\lambda \Rightarrow d = \frac{m\lambda}{\sin\theta_m} = \frac{(2)(600\times10^{-9}\text{ m})}{\sin(39.5°)} = 1.89\times10^{-6}\text{ m}$$

The number of lines in per millimeter is $(1\times10^{-3}\text{ m})/(1.89\times10^{-6}\text{ m}) = 530$.

22.13. Model: A diffraction grating produces an interference pattern.
Visualize: The interference pattern looks like the diagram in Figure 22.8.

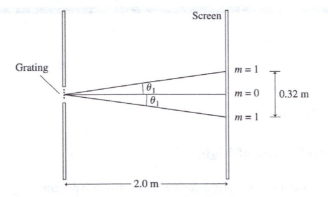

Solve: The bright fringes are given by Equation 22.15:

$$d\sin\theta_m = m\lambda \qquad m = 0, 1, 2, 3, \dots \Rightarrow d\sin\theta_1 = (1)\lambda \Rightarrow d = \lambda/\sin\theta_1$$

The angle θ_1 can be obtained from geometry as follows:

$$\tan\theta_1 = \frac{(0.32\text{ m})/2}{2.0\text{ m}} = 0.080 \Rightarrow \theta_1 = \tan^{-1}(0.080)\ 4.57°$$

Using $\sin\theta_1 = \sin 4.57° = 0.07968$,

$$d = \frac{633\times10^{-9}\text{ m}}{0.07968} = 7.9\ \mu\text{m}$$

Section 22.4 Single-Slit Diffraction

22.17. Visualize: We are given $L = 1.0\text{ m}$ and $w = 4000\lambda$.

Solve: Solve $w = \dfrac{2\lambda L}{a}$ for a.

$$a = \frac{2\lambda L}{w} = \frac{2\lambda(1.0\text{ m})}{4000\lambda} = \frac{1.0\text{ m}}{2000} = 0.50\text{ mm}$$

Assess: This is a typical slit width.

22.19. Model: A narrow slit produces a single-slit diffraction pattern.
Visualize: The intensity pattern for single-slit diffraction will look like Figure 22.14.
Solve: The width of the central maximum for a slit of width $a = 200\lambda$ is

$$w = \frac{2\lambda L}{a} = \frac{2(500\times10^{-9}\text{ m})(2.0\text{ m})}{0.0005\text{ m}} = 0.0040\text{ m} = 4.0\text{ mm}$$

Section 22.5 Circular-Aperture Diffraction

22.23. Model: Light passing through a circular aperture leads to a diffraction pattern that has a circular central maximum surrounded by a series of secondary bright fringes.

Visualize: The intensity pattern will look like Figure 22.15.
Solve: According to Equation 22.23, the angle that locates the first minimum in intensity is

$$\theta_1 = \frac{1.22\lambda}{D} = \frac{(1.22)(2.5\times10^{-6} \text{ m})}{0.20\times10^{-3} \text{ m}} = 0.01525 \text{ rad} = 0.874°$$

These should be rounded to $0.015 \text{ rad} = 0.87°$.

22.25. Model: Light passing through a circular aperture leads to a diffraction pattern that has a circular central maximum surrounded by a series of secondary bright fringes.
Visualize: The intensity pattern will look like Figure 22.15.
Solve: From Equation 22.24, the diameter of the circular aperture is

$$D = \frac{2.44\lambda L}{w} = \frac{2.44(633\times10^{-9} \text{ m})(4.0 \text{ m})}{2.5\times10^{-2} \text{ m}} = 0.25 \text{ mm}$$

Section 22.6 Interferometers

22.27. Model: An interferometer produces a new maximum each time L_2 increases by $\frac{1}{2}\lambda$ causing the path-length difference Δr to increase by λ.
Visualize: Please refer to the interferometer in Figure 22.20.
Solve: From Equation 22.33, the wavelength is

$$\lambda = \frac{2\Delta L_2}{\Delta m} = \frac{2(100\times10^{-6} \text{ m})}{500} = 4.0\times10^{-7} \text{ m} = 400 \text{ nm}$$

22.33. Solve: According to Equation 22.7, the fringe spacing between the m fringe and the $m + 1$ fringe is $\Delta y = \lambda L/d$. Δy can be obtained from Figure P22.33. The separation between the $m = 2$ fringes is 2.0 cm, implying that the separation between the two consecutive fringes is $\frac{1}{4}(2.0 \text{ cm}) = 0.50 \text{ cm}$. Thus,

$$\Delta y = 0.50\times10^{-2} \text{ m} = \frac{\lambda L}{d} \Rightarrow L = \frac{d\Delta y}{\lambda} = \frac{(0.20\times10^{-3} \text{ m})(0.50\times10^{-2} \text{ m})}{600\times10^{-9} \text{ m}} = 167 \text{ cm}$$

Assess: A distance of 167 cm from the slits to the screen is reasonable.

22.35. Solve: The intensity of light of a double-slit interference pattern at a position y on the screen is

$$I_{\text{double}} = 4I_1\cos^2\left(\frac{\pi d}{\lambda L}y\right)$$

where I_1 is the intensity of the light from each slit alone. At the center of the screen, that is, at $y = 0$ m, $I_1 = \frac{1}{4}I_{\text{double}}$. From Figure P22.33, I_{double} at the central maximum is 12 mW/m^2. So, the intensity due to a single slit is $I_1 = 3 \text{ mW/m}^2$.

22.39. Model: Each wavelength of light is diffracted at a different angle by a diffraction grating.
Solve: Light with a wavelength of 501.5 nm creates a first-order fringe at $y = 21.90$ cm. This light is diffracted at angle

$$\theta_1 = \tan^{-1}\left(\frac{21.90 \text{ cm}}{50.00 \text{ cm}}\right) = 23.65°$$

We can then use the diffraction equation $d\sin\theta_m = m\lambda$, with $m = 1$, to find the slit spacing:

$$d = \frac{\lambda}{\sin\theta_1} = \frac{501.5 \text{ nm}}{\sin(23.65°)} = 1250 \text{ nm}$$

The unknown wavelength creates a first order fringe at $y = 31.60$ cm, or at angle

$$\theta_1 = \tan^{-1}\left(\frac{31.60 \text{ cm}}{50.00 \text{ cm}}\right) = 32.29°$$

With the split spacing now known, we find that the wavelength is

$$\lambda = d\sin\theta_1 = (1250 \text{ nm})\sin(32.29°) = 667.8 \text{ nm}$$

Assess: The distances to the fringes and the first wavelength were given to 4 significant figures. Consequently, we can determine the unknown wavelength to 4 significant figures.

22.41. Model: We will assume that the listeners did not hear any other loud spots between the center and 1.4 m on each side, $m = 1$. We'll use the diffraction grating equations. First solve Equation 22.16 for θ_1 and insert it into Equation 22.15.

We need the wavelength of the sound waves, and we'll use the fundamental relationship for periodic waves to get it.

$$\lambda = \frac{v}{f} = \frac{340 \text{ m/s}}{10000 \text{ Hz}} = 3.4 \text{ cm} = 0.034 \text{ m}$$

We are also given $L = 10$ m and $y_1 = 1.4$ m.

Solve:

$$\theta_1 = \tan^{-1}\left(\frac{y_1}{L}\right) = \tan^{-1}\left(\frac{1.4 \text{ m}}{10 \text{ m}}\right) = 0.14 \text{ rad}$$

(This may be small enough to use the small-angle approximation, but we are almost finished with the problem without it, so maybe we can save the approximation and try it in the assess step.)

$$d = \frac{(1)\lambda}{\sin\theta_1} = \frac{0.034 \text{ m}}{\sin(0.14 \text{ rad})} = 0.25 \text{ m} = 25 \text{ cm}$$

Assess: These numbers seem reasonable given the size (wavelength) of sound waves.
With the small-angle approximation, $\theta \approx \sin\theta \approx \tan\theta$,

$$d = \frac{m\lambda}{y_m/L} = \frac{1(0.034 \text{ m})}{1.4 \text{ m}/10 \text{ m}} = 24 \text{ cm}$$

This is almost the same, but rounded down to two significant figures rather than up. The θ_1 we computed seemed almost small enough to use the approximation, and it would probably be OK for many applications when the angle is this small.

22.45. Model: A diffraction grating produces an interference pattern that is determined by both the slit spacing and the wavelength used.
Solve: (a) If blue light (the shortest wavelengths) is diffracted at angle θ, then red light (the longest wavelengths) is diffracted at angle $\theta + 30°$. In the first order, the equations for the blue and red wavelengths are

$$\sin\theta = \frac{\lambda_b}{d} \qquad d\sin(\theta + 30°) = \lambda_r$$

Combining the two equations we get for the red wavelength,

$$\lambda_r = d(\sin\theta\cos 30° + \cos\theta\sin 30°) = d(0.8660\sin\theta + 0.50\cos\theta) = d\left(\frac{\lambda_b}{d}\right)0.8660 + d(0.50)\sqrt{\left(1 - \frac{\lambda_b^2}{d^2}\right)}$$

$$\Rightarrow (0.50)d\sqrt{1 - \frac{\lambda_b^2}{d^2}} = \lambda_r - 0.8660\lambda_b \Rightarrow (0.50)^2(d^2 - \lambda_b^2) = (\lambda_r - 0.8660\lambda_b)^2$$

$$\Rightarrow d = \sqrt{\left(\frac{\lambda_r - 0.8660\lambda_b}{0.50}\right)^2 + \lambda_b^2}$$

Using $\lambda_b = 400\times10^{-9}$ m and $\lambda_r = 700\times10^{-9}$ m, we get $d = 8.125\times10^{-7}$ m. This value of d corresponds to

$$\frac{1 \text{ mm}}{d} = \frac{1.0\times10^{-3} \text{ m}}{8.125\times10^{-7} \text{ m}} = 1230 \text{ lines/mm}$$

(b) Using the value of d from part (a) and $\lambda = 589 \times 10^{-9}$ m, we can calculate the angle of diffraction as follows:
$$d \sin \theta_1 = (1)\lambda \Rightarrow (8.125 \times 10^{-7} \text{ m}) \sin \theta_1 = 589 \times 10^{-9} \text{ m} \Rightarrow \theta_1 = 46.5°$$

22.47. Model: A diffraction grating produces an interference pattern that is determined by both the slit spacing and the wavelength used.
Solve: From Figure P22.46,
$$\tan \theta_1 = \frac{0.436 \text{ m}}{1.0 \text{ m}} = 0.436 \Rightarrow \theta_1 = 23.557° \Rightarrow \sin \theta_1 = 0.400$$
Using the constructive-interference condition $d \sin \theta_m = m\lambda$,
$$d \sin 23.557° = (1)(600 \times 10^{-9} \text{ m}) \Rightarrow d = \frac{600 \times 10^{-9} \text{ m}}{\sin(23.557°)} = 1.50 \times 10^{-6} \text{ m}$$
Thus, the number of lines per millimeter is
$$\frac{1.0 \times 10^{-3} \text{ m}}{1.50 \times 10^{-6} \text{ m}} = 670 \text{ lines/mm}$$
Assess: The same answer is obtained if we perform calculations using information about the second-order bright constructive-interference fringe.

22.51. Model: Assume the incident light is coherent and monochromatic.
Visualize: The distances given in the table are $2y_1$ since the measurement is between the two first order fringes. Also note that $m = 1$ in this problem.
Solve: The equation for diffraction gratings is $d \sin \theta_m = m\lambda$ where d is the distance (in mm) between slits; we seek $\frac{1}{d}$, the number of lines per mm.
$$\sin \theta_1 = \frac{1}{d}\lambda$$
From $y_m = L \tan \theta_m$ and $y_1 = \frac{\text{distance}}{2}$, where "distance" is the distance between first order fringes as given in the table, we arrive at
$$\sin\left(\tan^{-1}\left(\frac{\text{distance}}{2L}\right)\right) = \frac{1}{d}\lambda$$
This leads us to believe that a graph of $\sin\left(\tan^{-1}\left(\frac{\text{distance}}{2L}\right)\right)$ vs. λ will produce a straight line whose slope is $\frac{1}{d}$ and whose intercept is zero.

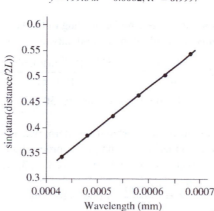

Line density
$y = 799.84x - 0.0002, R^2 = 0.9997$

We see from the spreadsheet that the linear fit is excellent and that the slope is 799.84 mm^{-1} and the intercept is very small. Because the number of lines per mm is always reported as an integer we round this answer to 800 lines/mm. **Assess:** This number of lines per mm is a typical number for a decent grating.

22.53. Model: A narrow slit produces a single-slit diffraction pattern.
Visualize: The diffraction-intensity pattern from a single slit will look like Figure 22.14.
Solve: These are not small angles, so we can't use equations based on the small-angle approximation. As given by Equation 22.19, the dark fringes in the pattern are located at $a \sin \theta_p = p\lambda$, where $p = 1, 2, 3, \ldots$ For the first minimum of the pattern, $p = 1$. Thus,

$$\frac{a}{\lambda} = \frac{p}{\sin \theta_p} = \frac{1}{\sin \theta_1}$$

For the three given angles the slit width to wavelength ratios are

(a): $\left(\dfrac{a}{\lambda}\right)_{30°} = \dfrac{1}{\sin 30°} = 2,$ **(b):** $\left(\dfrac{a}{\lambda}\right)_{60°} = \dfrac{1}{\sin 60°} = 1.15,$ **(c):** $\left(\dfrac{a}{\lambda}\right)_{90°} = \dfrac{1}{\sin 90°} = 1$

Assess: It is clear that the smaller the a/λ ratio, the wider the diffraction pattern. This is a conclusion that is contrary to what one might expect.

22.55. Model: A narrow slit produces a single-slit diffraction pattern.
Visualize: The dark fringes in this diffraction pattern are given by Equation 22.21:

$$y_p = \frac{p\lambda L}{a} \qquad p = 1, 2, 3, \ldots$$

We note that the first minimum in the figure is 0.50 cm away from the central maximum. We are given $a = 0.02$ nm and $L = 1.5$ m.
Solve: Solve the above equation for λ.

$$\lambda = \frac{y_p a}{pL} = \frac{(0.50 \times 10^{-2} \text{ m})(0.20 \times 10^{-3} \text{ m})}{(1)(1.5 \text{ m})} = 670 \text{ nm}$$

Assess: 670 nm is in the visible range.

22.59. Model: Light passing through a circular aperture leads to a diffraction pattern that has a circular central maximum surrounded by a series of secondary bright fringes.
Solve: (a) Because the visible spectrum spans wavelengths from 400 nm to 700 nm, we take the average wavelength of sunlight to be 550 nm.
(b) Within the small-angle approximation, the width of the central maximum is

$$w = 2.44 \frac{\lambda L}{D} \Rightarrow (1 \times 10^{-2} \text{ m}) = (2.44) \frac{(550 \times 10^{-9} \text{ m})(3 \text{ m})}{D} \Rightarrow D = 4.03 \times 10^{-4} \text{ m} = 0.40 \text{ mm}$$

22.61. Model: The laser light is diffracted by the circular opening of the laser from which the beam emerges.
Solve: The diameter of the laser beam is the width of the central maximum. We have

$$w = \frac{2.44 \lambda L}{D} \Rightarrow D = \frac{2.44 \lambda L}{w} = \frac{2.44(532 \times 10^{-9} \text{ m})(3.84 \times 10^{8} \text{ m})}{1000 \text{ m}} = 0.50 \text{ m}$$

In other words, the laser beam must emerge from a laser of diameter 50 cm.

22.63. Model: A diffraction grating produces an interference pattern, which looks like the diagram of Figure 22.8.
Solve: (a) Nothing has changed while the aquarium is empty. The order of a bright (constructive interference) fringe is related to the diffraction angle θ_m by $d \sin \theta_m = m\lambda$, where $m = 0, 1, 2, 3, \ldots$ The space between the slits is

$$d = \frac{1.0 \text{ mm}}{600} = 1.6667 \times 10^{-6} \text{ m}$$

For $m=1,$

$$\sin\theta_1 = \frac{\lambda}{d} \Rightarrow \theta_1 = \sin^{-1}\left(\frac{633\times10^{-9}\ \text{m}}{1.6667\times10^{-6}\ \text{m}}\right) = 22.3°$$

(b) The path-difference between the waves that leads to constructive interference is an integral multiple of the wavelength in the medium in which the waves are traveling, that is, water. Thus,

$$\lambda = \frac{633\ \text{nm}}{n_{\text{water}}} = \frac{633\ \text{nm}}{1.33} = 4.759\times10^{-7}\ \text{m} \Rightarrow \sin\theta_1 = \frac{\lambda}{d} = \frac{4.759\times10^{-7}\ \text{m}}{1.6667\times10^{-6}\ \text{m}} = 0.2855 \Rightarrow \theta_1 = 16.6°$$

22.67. Model: The arms of the interferometer are of equal length, so without the crystal the output would be bright.
Visualize: We need to consider how many more wavelengths fit in the electro-optic crystal than would have occupied that space (6.70 μm) without the crystal; if it is an integer then the interferometer will produce a bright output; if it is a half-integer then the interferometer will produce a dark output. But the wavelength we need to consider is the wavelength inside the crystal, not the wavelength in air.

$$\lambda_n = \frac{\lambda}{n}$$

We are told the initial n with no applied voltage is 1.522, and the wavelength in air is $\lambda = 1.000\ \mu$m.
Solve: The number of wavelengths that would have been in that space without the crystal is

$$\frac{6.70\ \mu\text{m}}{1.000\ \mu\text{m}} = 6.70$$

(a) With the crystal in place (and $n = 1.522$) the number of wavelengths in the crystal is

$$\frac{6.70\ \mu\text{m}}{1.000\ \mu\text{m}/1.522} = 10.20$$

$$10.20 - 6.70 = 3.50$$

which shows there are a half-integer number more wavelengths with the crystal in place than if it weren't there. Consequently the output is dark with the crystal in place but no applied voltage.
(b) Since the output was dark in the previous part, we want it to be bright in the new case with the voltage on. That means we want to have just one-half more extra wavelengths in the crystal (than if it weren't there) than we did in the previous part. That is, we want 4.00 extra wavelengths in the crystal instead of 3.5, so we want $6.70 + 4.00 = 10.70$ wavelengths in the crystal.

$$\frac{6.70\ \mu\text{m}}{1.000\ \mu\text{m}/n} = 10.70 \qquad \Rightarrow \qquad n = \frac{10.70(1.000\ \mu\text{m})}{6.70\ \mu\text{m}} = 1.597$$

Assess: It seems reasonable to be able to change the index of refraction of a crystal from 1.522 to 1.597 by applying a voltage.

RAY OPTICS

Exercises and Problems

Section 23.1 The Ray Model of Light

23.1. Model: Light rays travel in straight lines.
Solve: (a) The time is

$$t = \frac{\Delta x}{c} = \frac{1.0 \text{ m}}{3.0 \times 10^8 \text{ m/s}} = 3.3 \times 10^{-9} \text{ s} = 3.3 \text{ ns}$$

(b) The refractive indices for water, glass, and cubic zirconia are 1.33, 1.50, and 1.96, respectively. In a time of 3.33 ns, light will travel the following distance in water:

$$\Delta x_{\text{water}} = v_{\text{water}} t = \left(\frac{c}{n_{\text{water}}} \right) t = \left(\frac{3.0 \times 10^8 \text{ m/s}}{1.33} \right) \left(3.33 \times 10^{-9} \text{ s} \right) = 0.75 \text{ m}$$

Likewise, the distances traveled in the glass and cubic zirconia are $\Delta x_{\text{glass}} = 0.67$ m and $\Delta x_{\text{cubic zirconia}} = 0.46$ m.

Assess: The higher the refractive index of a medium, the slower the speed of light and hence smaller the distance it travels in that medium in a given time.

Section 23.2 Reflection

23.5. Model: Use the ray model of light.
Visualize:

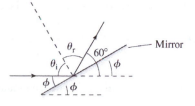

According to the law of reflection, $\theta_r = \theta_i$.
Solve: From the geometry of the diagram,

$$\theta_i + \phi = 90° \qquad \theta_r + (60° - \phi) = 90°$$

Using the law of reflection, we get

$$90° - \phi = 90° - (60° - \phi) \quad \Rightarrow \quad \phi = 30°$$

Assess: The above result leads to a general result for plane mirrors: If a plane mirror makes an angle ϕ relative to the incident ray, the reflected ray makes an angle of 2ϕ with respect to the incident ray.

23.9. Model: Use the ray model of light and the law of reflection.
Visualize:

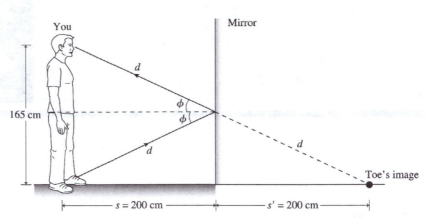

We only need one ray of light that leaves your toes and reflects into your eye.
Solve: From the geometry of the diagram, the distance from your eye to your toes' image is

$$2d = \sqrt{(400 \text{ cm})^2 + (165 \text{ cm})^2} = 433 \text{ cm}$$

Assess: The light appears to come from your toes' image.

Section 23.3 Refraction

23.11. Model: Use the ray model of light and Snell's law.
Visualize: See the figure below. Note that the angle we need to find is θ_2.

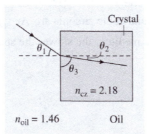

Solve: To find θ_2, apply Snell's law at the interface between the oil and the cubic zirconium:

$$n_{\text{oil}} \sin \theta_1 = n_{CZ} \sin \theta_2 \quad \Rightarrow \quad \theta_2 = \sin^{-1}\left(\frac{n_{\text{oil}}}{n_{CZ}} \sin \theta_1\right) = \sin^{-1}\left(\frac{1.46}{2.18} \sin 25°\right) = 16°$$

23.15. Model: Use the ray model of light. For an angle of incidence greater than the critical angle, the ray of light undergoes total internal reflection.
Visualize:

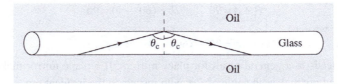

Solve: The critical angle of incidence is given by Equation 23.9:

$$\theta_c = \sin^{-1}\left(\frac{n_{oil}}{n_{glass}}\right) = \sin^{-1}\left(\frac{1.46}{1.50}\right) = 76.7°$$

Assess: The critical angle exists because $n_{oil} < n_{glass}$.

Section 23.4 Image Formation by Refraction

23.17. Model: Represent the beetle as a point source and use the ray model of light.
Visualize:

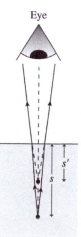

Eye

Paraxial rays from the beetle refract into the air and then enter into the observer's eye. The rays in the air when extended into the plastic appear to be coming from the beetle at a shallower location, a distance s' from the plastic-air boundary.
Solve: The actual object distance is s and the image distance is $s' = 2.0$ cm. Using Equation 23.13,

$$s' = \frac{n_2}{n_1}s = \frac{n_{air}}{n_{plastic}}s \quad \Rightarrow \quad 2.0 \text{ cm} = \frac{1.0}{1.59}s \quad \Rightarrow \quad s = 3.2 \text{ cm}$$

Assess: The beetle is much deeper in the plastic than it appears to be.

Section 23.5 Color and Dispersion

23.21. Model: Use the ray model of light and the phenomenon of dispersion.
Visualize:

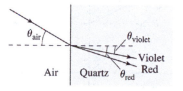

Solve: Using Snell's law for the red light,

$$n_{air}\sin\theta_{air} = n_{red}\sin\theta_{red} \quad \Rightarrow \quad (1.00)\sin\theta_{air} = 1.45\sin(26.3°) \quad \Rightarrow \quad \theta_{air} = \frac{\sin^{-1}(1.45\sin 26.3°)}{1.00} = 40.0°$$

Now using Snell's law for the violet light,

$$n_{air}\sin\theta_{air} = n_{violet}\sin\theta_{violet} \quad \Rightarrow \quad (1.00)\sin 40.0° = n_{violet}\sin 25.7° \quad \Rightarrow \quad n_{violet} = 1.48$$

Assess: As expected, n_{violet} is slightly larger than n_{red}.

23.23. Model: The intensity of scattered light is inversely proportional to the fourth power of the wavelength.

Solve: We want to find the wavelength of infrared light such that $I_{IR} = 0.01 I_{500}$. Because $I_{500} \propto (500 \text{ nm})^{-4}$ and $I_{IR} \propto \lambda^{-4}$, we have

$$\frac{I_{500}}{I_{IR}} = \left(\frac{\lambda}{500 \text{ nm}} \right)^4 = 100 \quad \Rightarrow \quad \lambda = 1580 \text{ nm} \approx 1600 \text{ nm}$$

Section 23.6 Thin Lenses: Ray Tracing

23.25. Model: Use ray tracing to locate the image.
Solve:

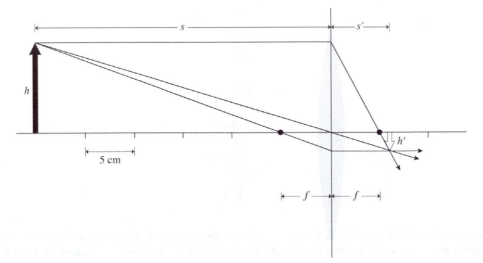

The figure shows the ray-tracing diagram using the steps of Tactics Box 23.2. You can see from the diagram that the image is in the plane where the three special rays converge. The image is located at $s' = 6.0$ cm behind the converging lens, and is inverted.

23.27. Model: Use ray tracing to locate the image.
Solve:

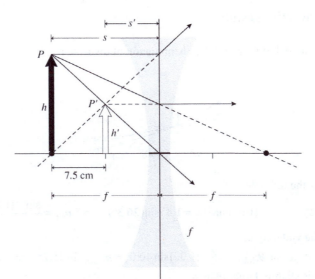

The figure shows the ray-tracing diagram using the steps of Tactics Box 23.3. The three rays after refraction do not converge at a point, but they appear to come from P′. P′ is 7.5 cm from the diverging lens, so $s' = -7.5$ cm. Thus, the image is in front of the lens and is upright.

Section 23.7 Thin Lenses: Refraction Theory

23.31. Model: Assume the meniscus lens is a thin lens.
Solve: If the object is on the left, then the first surface has $R_1 = 30$ cm (convex toward the object) and the second surface has $R_2 = 40$ cm (convex toward the object). The index of refraction of polystyrene plastic is 1.59, so the lensmaker's equation gives

$$\frac{1}{f} = (n-1)\left(\frac{1}{R_1} - \frac{1}{R_2}\right) = (1.59-1)\left(\frac{1}{30\text{ cm}} - \frac{1}{40\text{ cm}}\right) \quad \Rightarrow \quad f = 203\text{ cm} = 0.20\text{ m}$$

Section 23.8 Image Formation with Spherical Mirrors

23.35. Solve: The image is at 40 cm in front of the mirror and is inverted.

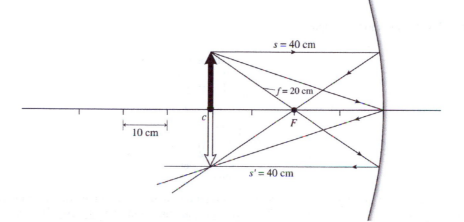

Assess: When the object is outside the focal length we get an inverted image.

23.37. Solve: The image is at −12 cm; that is, it is 12 cm behind the mirror and is upright.

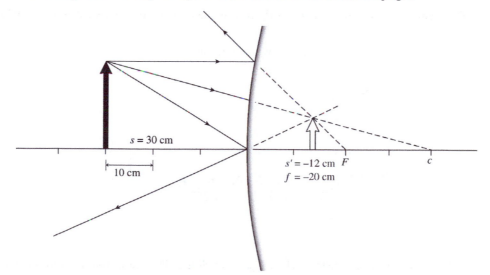

Assess: We expected an upright virtual image from the convex mirror.

23.39. Model: Treat the red ball as a point source and use the ray model of light.
Solve: **(a)** Using the law of reflection, we can obtain 3 images of the red ball.
(b) The images of the ball are located at B, C, and D. Relative to the intersection point of the two mirrors, the coordinates of B, C, and D are B(+1.0 m, −2.0 m), C(−1.0 m, +2.0 m), and D(+1.0 m, +2.0 m).
(c)

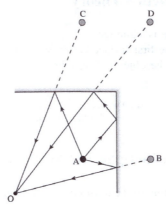

23.43. Model: Use the ray model of light.
Visualize:

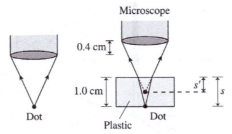

Solve: When the plastic is in place, the microscope focuses on the virtual image of the dot. From the figure, we note that $s = 1.0$ cm and $s' = 1.0$ cm $− 0.4$ cm $= 0.6$ cm. The rays are paraxial, and the object and image distances are measured relative to the plastic-air boundary. Using Equation 23.13,

$$s' = \frac{n_{air}}{n_{plastic}} s \quad \Rightarrow \quad 0.60 \text{ cm} = \frac{1.00}{n_{plastic}}(1.00 \text{ cm}) \quad \Rightarrow \quad n_{plastic} = \frac{1.00 \text{ cm}}{0.60 \text{ cm}} = 1.7$$

Assess: This value seems reasonable because it is fairly close to the value for polystyrene plastic $(n = 1.59)$.

23.45. Model: Use the ray model of light and the law of refraction.
Visualize:

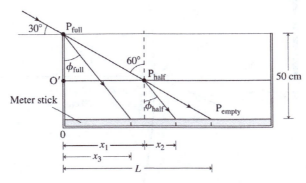

Solve: **(a)** The ray of light strikes the meter stick at P_{empty}, which is a distance L from the zero mark of the meter stick. So,

$$\tan 60° = \frac{L}{50 \text{ cm}} \quad \Rightarrow \quad L = (50 \text{ cm})\tan 60° = 87 \text{ cm}$$

(b) The ray of light refracts at P_{half} and strikes the meter stick a distance $x_1 + x_2$ from the zero of the meter stick. We can find x_1 from the triangle $P_{\text{full}}P_{\text{half}}O'$:

$$\tan 60° = \frac{x_1}{25 \text{ cm}} \quad \Rightarrow \quad x_1 = (25 \text{ cm})\tan 60° = 43.30 \text{ cm}$$

We also have $x_2 = (25 \text{ cm})\tan \phi_{\text{half}}$. Using Snell's law,

$$n_{\text{air}} \sin 60° = n_{\text{water}} \sin \phi_{\text{half}} \quad \Rightarrow \quad \phi_{\text{half}} = \sin^{-1}\left(\frac{\sin 60°}{1.33}\right) = 40.63°$$

$$x_2 = (25 \text{ cm})\tan 40.63° = 21.45 \text{ cm} \quad \Rightarrow \quad x_1 + x_2 = 43.30 \text{ cm} + 21.45 \text{ cm} = 65 \text{ cm}$$

(c) The ray of light experiences refraction at P_{full} and the angle of refraction is the same as in part (b). We get

$$\tan \phi_{\text{full}} = \frac{x_3}{50 \text{ cm}} \quad \Rightarrow \quad x_3 = (50 \text{ cm})\tan 40.63° = 43 \text{ cm}$$

23.49. Model: Use the ray model of light and the law of refraction. Assume that the laser beam is a ray of light.
Visualize:

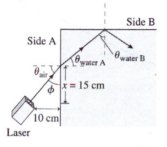

Solve: (a) From the geometry of the diagram at side A, we have

$$\tan \phi = \frac{10 \text{ cm}}{15 \text{ cm}} \quad \Rightarrow \quad \phi = \tan^{-1}\left(\frac{10}{15}\right) \quad \Rightarrow \quad \phi = 33.69°$$

This means the angle of incidence at side A is $\theta_{\text{air}} = 90° - 33.69° = 56.31°$. Using Snell's law at side A gives

$$n_{\text{air}} \sin \theta_{\text{air}} = n_{\text{water}} \sin \theta_{\text{water A}} \quad \Rightarrow \quad \theta_{\text{water A}} = \sin^{-1}\left(\frac{1.00\sin(56.31°)}{1.33}\right) = 38.73°$$

This ray of light now strikes side B. The angle of incidence at this water-air boundary is $\theta_{\text{water B}} = 90° - \theta_{\text{water A}} = 51.3°$. The critical angle for the water-air boundary is

$$\theta_{\text{c}} = \sin^{-1}\left(\frac{n_{\text{air}}}{n_{\text{water}}}\right) = \sin^{-1}\left(\frac{1.0}{1.33}\right) = 48.8°$$

Because the angle $\theta_{\text{water B}} > \theta_{\text{c}}$, the ray will experience total internal reflection.
(b) We will now repeat the above calculation with $x = 25$ cm. From the geometry of the diagram at side A, $\phi = 21.80°$ and $\theta_{\text{air}} = 68.20°$. Using Snell's law at the air-water boundary, $\theta_{\text{water A}} = 44.28°$ and $\theta_{\text{water B}} = 45.72°$. Because $\theta_{\text{water B}} < \theta_{\text{c}}$, the ray will be refracted into the air. The angle of refraction is calculated as follows:

$$n_{\text{air}} \sin \theta_{\text{air B}} = n_{\text{water}} \sin \theta_{\text{water B}} \quad \Rightarrow \quad \theta_{\text{air B}} = \sin^{-1}\left(\frac{1.33\sin(45.72°)}{1.00}\right) = 72°$$

(c) Using the critical angle for the water-air boundary found in part (a), $\theta_{\text{water A}} = 90° - 48.75° = 41.25°$. According to Snell's law,

$$n_{\text{air}} \sin \theta_{\text{air}} = n_{\text{water}} \sin \theta_{\text{water A}} \Rightarrow \theta_{\text{air}} = \sin^{-1}\left(\frac{1.33 \sin 41.25°}{1.0}\right) = 61.27°$$

$$\phi = 90° - 61.27° = 28.73°$$

The minimum value of x for which the laser beam passes through side B and emerges into the air is calculated as follows:

$$\tan \phi = \frac{10 \text{ cm}}{x} \quad \Rightarrow \quad x = \frac{10 \text{ cm}}{\tan 28.73°} = 18.2 \text{ cm} \approx 18 \text{ cm}$$

23.51. Model: Use the ray model of light and the law of refraction. Assume that the laser beam is a ray of light.
Solve: Applying Snell's Law (Equation 23.3) to the data gives

$$n_{\text{air}} \sin \theta_i = n_{\text{plastic}} \sin \theta_r \quad \Rightarrow \quad \sin \theta_i = \frac{n_{\text{plastic}}}{1.00} \sin \theta_r$$

where we have used $n_{\text{air}} = 1.00$. Thus, if we plot $\sin \theta_i$ versus $\sin \theta_r$, the slope of the curve is n_{plastic}.

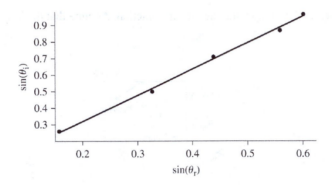

The slope of the fit to the data gives $n_{\text{plastic}} = 1.58$.
Assess: The result is reasonable for plastic.

23.55. Model: Use the ray model of light. The surface is a spherically refracting surface.
Visualize:

Solve: Because the rays are parallel, $s = \infty$. The rays come to focus on the rear surface of the sphere, so $s' = 2R$, where R is the radius of curvature of the sphere. Equation 23.21 gives

$$\frac{n_1}{s} + \frac{n_2}{s'} = \frac{n_2 - n_1}{R} \quad \Rightarrow \quad \frac{1}{\infty} + \frac{n}{2R} = \frac{n-1}{R} \quad \Rightarrow \quad n = 2.00$$

23.57. Model: Use ray tracing to locate the image. Assume that the converging lens is a thin lens.
Solve: **(a)**

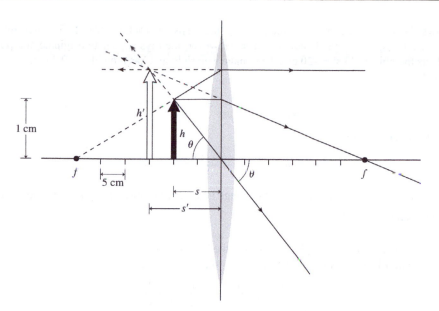

The figure shows the ray-tracing diagram made using the steps of Tactics Box 23.2. The three special rays that experience refraction do not converge at a point. Instead they appear to come from a point that is 15 cm on the same side as the object itself. Thus $s' = -15$ cm. The image is upright and has a height of $h' = 1.5$ cm.
(b) Using the thin-lens formula,

$$\frac{1}{s} + \frac{1}{s'} = \frac{1}{f} \quad \Rightarrow \quad \frac{1}{10 \text{ cm}} + \frac{1}{s'} = \frac{1}{30 \text{ cm}} \quad \Rightarrow \quad \frac{1}{s'} = -\frac{1}{15 \text{ m}} \quad \Rightarrow \quad s' = -15 \text{ cm}$$

The image height is obtained from

$$m = -\frac{s'}{s} = -\frac{-15 \text{ cm}}{10 \text{ cm}} = +1.5$$

The image is upright and 1.5 times the object, that is, 1.5 cm high. These values agree with those obtained in part (a).

23.61. Model: Use ray tracing to locate the image. Assume the diverging lens is a thin lens.
Solve: (a)

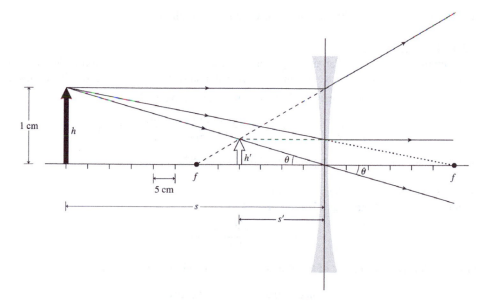

The figure shows the ray-tracing diagram made using the steps of Tactics Box 23.3. After refraction from the diverging lens, the three special rays do not converge. However, the rays appear to originate at a point that is 20 cm on the same side as the object. So $s' = -20$ cm. The image is upright and has a height of 0.3 cm.
(b) The thin-lens formula gives

$$\frac{1}{s'} = \frac{1}{f} - \frac{1}{s} = \frac{1}{-30 \text{ cm}} - \frac{1}{60 \text{ cm}} = -\frac{1}{20 \text{ cm}} \quad \Rightarrow \quad s' = -20 \text{ cm}$$

The image height is obtained from

$$m = -\frac{s'}{s} = -\frac{-20 \text{ cm}}{60 \text{ cm}} = \frac{1}{3} = 0.33$$

Thus, $h' = |m|h = (0.33)(1.0 \text{ cm}) = 0.33 \text{ cm},$ and the image is upright because m is positive. These values for s' and h' agree with those obtained in part (a).

23.63. Model: Use the ray model of light and assume the lens is thin.
Solve: Applying the thin-lens equation (Equation 23.26) to the data, we find

$$\frac{1}{s} + \frac{1}{s'} = \frac{1}{f} \quad \Rightarrow \quad \frac{1}{s} = \frac{1}{f} - \frac{1}{s'}$$

So if we plot the inverse object distance as a function of the inverse image distance, the y-intercept will be the inverse focal length.

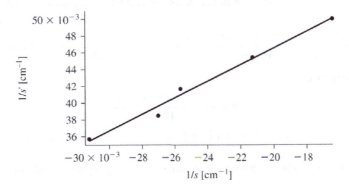

The fit gives the y-intercept at 0.066259 cm^{-1}, which gives a focal length of $f = 15.1$ cm

23.67. Visualize: Refer to Figure 23.57.
Solve: We are given that $s = +1.2$ cm. For an upright image, $m > 0$, so $m = +1.5$. Equation 23.30 then gives the image distance: $s' = -ms = -(1.5)(1.2 \text{ cm}) = -1.8$ cm. This is a virtual image that is located on the opposite side of the mirror from the object. The focal length may be found by using Equation 23.28:

$$\frac{1}{s} + \frac{1}{s'} = \frac{1}{f} \quad \Rightarrow \quad f = \frac{ss'}{s+s'} = \frac{(1.2 \text{ cm})(-1.8 \text{ cm})}{1.2 \text{ cm} - 1.8 \text{ cm}} = 3.6 \text{ cm}$$

The focal length is positive, so the mirror is concave toward the object.

23.71. Model: Assume the projector lens is a thin lens.
Solve: (a) The absolute value of the magnification of the lens is

$$|m| = \left|\frac{h'}{h}\right| = \left|\frac{98 \text{ cm}}{2.0 \text{ cm}}\right| = 49$$

Because the projector forms a real image of a real object, the image will be inverted. Thus,

$$m = -49 = -\frac{s'}{s} \quad \Rightarrow \quad s' = 49s$$

We also have

$$s + s' = 300 \text{ cm} \implies s + 49s = 300 \text{ cm} \implies s = 6.0 \text{ cm} \implies s' = 294 \text{ cm}$$

Using these values of s and s', we can find the focal length of the lens:

$$\frac{1}{f} = \frac{1}{s} + \frac{1}{s'} = \frac{1}{6.0 \text{ cm}} + \frac{1}{294 \text{ cm}} \implies f = 5.9 \text{ cm}$$

(b) From part (a) the lens should be 6.0 cm from the slide.

23.73. Model: Assume the symmetric converging lens is a thin lens.
Solve: Because the lens forms a real image on the screen of a real object, the image is inverted. Thus, $m = -2 = -s'/s$. Also,

$$s + s' = 60 \text{ cm} \implies s + 2s = 60 \text{ cm} \implies s = 20 \text{ cm} \implies s' = 40 \text{ cm}$$

We can use the thin-lens formula to determine the radius of curvature of the symmetric converging lens ($|R_1| = |R_2|$) as follows:

$$\frac{1}{s} + \frac{1}{s'} = \frac{1}{f} = (n-1)\left(\frac{1}{R_1} - \frac{1}{R_2}\right)$$

Using $R_1 = +|R|$ (convex toward the object), $R_2 = -|R|$ (concave toward the object), and $n = 1.59$, we find

$$\frac{1}{20 \text{ cm}} + \frac{1}{40 \text{ cm}} = (1.59-1)\left(\frac{1}{|R|} - \frac{1}{-|R|}\right) \implies \frac{3}{40 \text{ cm}} = \frac{1.18}{|R|} \implies |R| = 15.7 \text{ cm} \approx 16 \text{ cm}$$

23.77. Visualize: First concentrate on the optic axis and the ray parallel to it. Geometry says if parallel lines are both cut by a diagonal (in this case the line through the center of curvature and normal to the mirror at the point of incidence) the interior angles are equal; so $\phi = \theta_i$. The law of reflection says that $\theta_i = \theta_r$, so we conclude $\phi = \theta_r$.
Now concentrate on the triangle whose sides are R, a, and b. Because two of the angles are equal then it is isosceles; therefore $b = a$. Apply the law of cosines to this triangle.
Solve:

$$b^2 = a^2 + R^2 - 2aR\cos\phi$$

Because $a = b$, they drop out.

$$R^2 = 2aR\cos\phi \implies R = 2a\cos\phi$$

We want to know how big a is in terms of R, so solve for a.

$$a = \frac{R}{2\cos\phi}$$

If $\phi \ll 1$ then $\cos\phi \approx 1$, so in the limit of small ϕ, $a = R/2$, and then since $f = R - a$ it must also be that

$$f = \frac{R}{2}$$

Assess: Many textbooks forget to stress that $f = R/2$ only in the limit of small ϕ, *i.e.*, for paraxial rays.

OPTICAL INSTRUMENTS

Exercises and Problems

Section 24.1 Lenses in Combination

24.1. Model: Each lens is a thin lens. The image of the first lens is the object for the second lens.
Visualize:

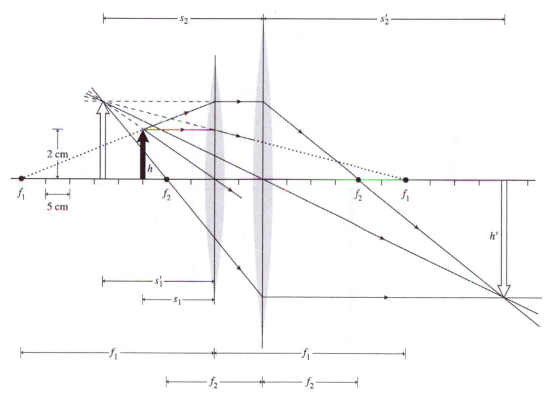

The figure shows the two lenses and a ray-tracing diagram. The ray-tracing shows that the lens combination will produce a real, inverted image behind the second lens.
Solve: (a) From the ray-tracing diagram, we find that the image is ≈ 50 cm from the second lens and the height of the final image is 4.5 cm.
(b) $s_1 = 15$ cm is the object distance of the first lens. Its image, which is a virtual image, is found from the thin-lens equation:

$$\frac{1}{s_1'} = \frac{1}{f_1} - \frac{1}{s_1} = \frac{1}{40\text{ cm}} - \frac{1}{15\text{ cm}} = -\frac{5}{120\text{ cm}} \Rightarrow s_1' = -24\text{ cm}$$

The magnification of the first lens is

$$m_1 = -\frac{s_1'}{s_1} = -\frac{(-24\text{ cm})}{15\text{ cm}} = 1.6$$

The image of the first lens is now the object for the second lens. The object distance is $s_2 = 24\text{ cm} + 10\text{ cm} = 34\text{ cm}$. A second application of the thin-lens equation yields:

$$\frac{1}{s_2'} = \frac{1}{f_2} - \frac{1}{s_2} = \frac{1}{20\text{ cm}} - \frac{1}{34\text{ cm}} \Rightarrow s_2' = \frac{680\text{ cm}}{14} = 48.6\text{ cm} \approx 49\text{ cm}$$

The magnification of the second lens is

$$m_2 = -\frac{s_2'}{s_2} = -\frac{48.6\text{ cm}}{34\text{ cm}} = -1.429$$

The combined magnification is $m = m_1 m_2 = (1.6)(-1.429) = -2.286$. The height of the final image is $(2.286)(2.0\text{ cm}) = 4.6\text{ cm}$. These calculated values are in agreement with those found in part (a).

24.3. Model: Each lens is a thin lens. The image of the first lens is the object for the second lens.
Visualize:

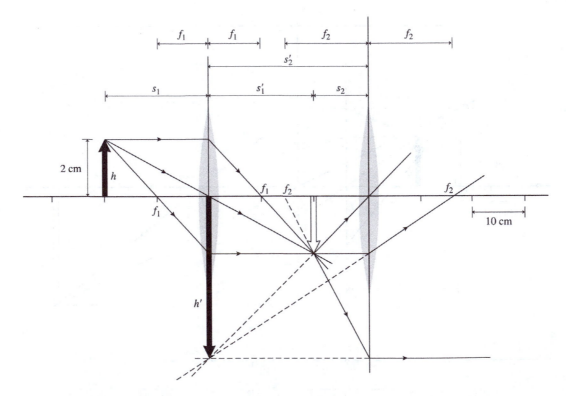

The figure shows the two lenses and a ray-tracing diagram. The ray tracing shows that the lens combination will produce a virtual, inverted image at the first lens.
Solve: (a) From the ray-tracing diagram, we find that the image is 30 cm in front of the second lens and the height of the final image is 6 cm.
(b) $s_1 = 20\text{ cm}$ is the object distance of the first lens. Its image, which is real and inverted, is found from the thin lens equation:

$$\frac{1}{s_1}+\frac{1}{s_1'}=\frac{1}{f_1}\Rightarrow s_1'=\frac{f_1 s_1}{s_1-f_1}=\frac{(10\text{ cm})(20\text{ cm})}{20\text{ cm}-10\text{ cm}}=20\text{ cm}$$

The magnification of the first lens is

$$m_1=-\frac{s_1'}{s_1}=-\frac{20\text{ cm}}{20\text{ cm}}=-1$$

The image of the first lens is now the object for the second lens. The object distance is $s_2=30\text{ cm}-20\text{ cm}=10\text{ cm}$. A second application of the thin lens equation yields

$$\frac{1}{s_2}+\frac{1}{s_2'}=\frac{1}{f_2}\Rightarrow s_2'=\frac{f_2 s_2}{s_2-f_2}=\frac{(15\text{ cm})(10\text{ cm})}{10\text{ cm}-15\text{ cm}}=-30\text{ cm}$$

The magnification of the second lens is

$$m_2=-\frac{s_2'}{s_2}=-\frac{(-30\text{ cm})}{10\text{ cm}}=3$$

The combined magnification is $m=m_1 m_2=(-1)(3)=-3$. The height of the final image is $(3)(2.0\text{ cm})=6.0\text{ cm}$. The image is inverted because m has a negative sign. These calculated values are in agreement with those found in part (a). **Assess:** The thin lens equation agrees with the ray tracing.

Section 24.2 The Camera

24.7. Visualize: Equation 24.2 gives $f\text{-number}=f/D$.
Solve:

$$f\text{-number}=\frac{f}{D}=\frac{35\text{ mm}}{7.0\text{ mm}}=5.0$$

Assess: This is in the range of f-numbers for typical camera lenses.

24.11. Visualize: We want the same exposure in both cases. The exposure depends on $I\Delta t_{\text{shutter}}$. We'll also use Equation 24.3. The lens is the same lens in both cases, so $f=f'$.
Solve:

$$\text{exposure}=I\Delta t\propto\frac{D^2}{f^2}\Delta t$$

$$\frac{D^2}{f^2}\Delta t=\frac{D'^2}{f'^2}\Delta t'$$

Solve for D'; then simplify.

$$D'=\sqrt{D^2\left(\frac{f'}{f}\right)^2\left(\frac{\Delta t}{\Delta t'}\right)}=D\sqrt{\frac{\Delta t}{\Delta t'}}=(3.0\text{ mm})\sqrt{\frac{1/125\text{ s}}{1/500\text{ s}}}=(3.0\text{ mm})\sqrt{4}=6.0\text{ mm}$$

Assess: Since we decreased the shutter speed by a factor of 4 we need to increase the aperture area by a factor of 4, and this means increase the diameter by a factor of 2.

Section 24.3 Vision

24.13. Model: Ignore the small space between the lens and the eye.
Visualize: Refer to Example 24.5, but we want to solve for s', the far point.
Solve:
(a) The power of the lens is negative which means the focal length is negative, so Ellen wears diverging lenses. This is the remedy for myopia.

(b) We want to know where the image should be for an object $s = \infty$ m given $1/f = -1.0$ m^{-1}.

$$f = \frac{1}{P} = -1.0 \text{ m}$$

$$\frac{1}{s} + \frac{1}{s'} = \frac{1}{f}$$

When $s = \infty$ m,

$$\frac{1}{\infty \text{ m}} + \frac{1}{s'} = \frac{1}{f} \Rightarrow s' = f = -1.0 \text{ m}$$

So the far point is 100 cm.

Assess: The negative sign on s' is expected because we need the image to be virtual.

Section 24.4 Optical Systems That Magnify

24.17. Visualize: Equation 24.10 relates the variables in question:

$$M = -\frac{L}{f_{obj}} \frac{25 \text{ cm}}{f_{eye}}$$

We are given $M = 100\times$, $L = 160$ mm, and $f_{obj} = 8.0$ cm.

Solve: Solve for f_{eye}.

$$f_{eye} = -\frac{L}{f_{obj}} \frac{25 \text{ cm}}{M} = -\frac{160 \text{ mm}}{8.0 \text{ mm}} \frac{25 \text{ cm}}{(-100)} = 5.0 \text{ cm}$$

Assess: This is the same f_{eye} as in the previous exercise.

24.19. Visualize: Figure 24.14 shows from similar triangles that for the eyepiece lens to collect all the light

$$\frac{D_{obj}}{f_{obj}} = \frac{D_{eye}}{f_{eye}}$$

We also see from Equation 24.11 that $M = -f_{obj}/f_{eye}$. We are given $M = -20$ and $D_{obj} = 12$ cm.

Solve:

$$D_{eye} = D_{obj} \frac{f_{eye}}{f_{obj}} = \frac{D_{obj}}{-M} = \frac{12 \text{ cm}}{20} = 0.60 \text{ cm} = 6.0 \text{ mm}$$

Assess: The answer is almost as wide as a dark-adapted eye.

Section 24.5 The Resolution of Optical Instruments

24.23. Visualize: We are given $h_1 = 1.0$ cm, $s_1 = 4.0$ cm, $f_1 = 5.0$ cm, and $f_2 = -8.0$ cm.

Solve: First compute the image from the first lens.

$$s_1' = \frac{f_1 s_1}{s_1 - f_1} = \frac{(5.0 \text{ cm})(4.0 \text{ cm})}{4.0 \text{ cm} - 5.0 \text{ cm}} = -20 \text{ cm}$$

$$h_1' = -h_1 \frac{s_1'}{s_1} = -(1.0 \text{ cm}) \frac{-20 \text{ cm}}{4.0 \text{ cm}} = 5.0 \text{ cm}$$

This is a virtual, upright image 20 cm to the left of the first lens.

The second lens is 12 cm to the right of the first one, so $s_2 = 32$ cm.

$$s_2' = \frac{f_2 s_2}{s_2 - f_2} = \frac{(-8.0 \text{ cm})(32 \text{ cm})}{32 \text{ cm} - (-8.0 \text{ cm})} = -6.4 \text{ cm}$$

$$h_2 = h_1' = 5.0 \text{ cm}$$

$$h_2' = -h_2 \frac{s_2'}{s_2} = -(5.0 \text{ cm}) \frac{-6.4 \text{ cm}}{32 \text{ cm}} = 1.0 \text{ cm}$$

This is a virtual, upright image 6.4 cm to the left of the second lens (5.6 cm to the right of the first lens). The image is 1.0 cm tall (the same size as the object).

Assess: Ray tracing confirms these results.

24.25. Visualize: Physically, the light rays can either go directly through the lens or they can reflect from the mirror and then go through the lens. We can consider the image from the lens alone and then consider the image from mirror becoming the object for the lens.

Solve:

(a) First case: the lens gets the subscript 1's and the mirror the 2's. The location of the image from the lens is

$$s_1' = \frac{f_1 s_1}{s_1 - f_1} = \frac{(10 \text{ cm})(5 \text{ cm})}{5 \text{ cm} - 10 \text{ cm}} = -10 \text{ cm}$$

The image is right at the mirror plane and a calculation for a mirror shows that when $s_2 = 0$ then $s_2' = 0$, too. So the final image is at the mirror, 10 cm to the left of the lens.

$$m = -\frac{s_1'}{s_1} = -\frac{-10 \text{ cm}}{5 \text{ cm}} = 2.0$$

so $h' = hm = (1.0 \text{ cm})(2.0) = 2.0 \text{ cm}$.

Second case: the mirror gets the subscript 1's and the lens the 2's. The location of the image from the mirror is

$$s_1' = \frac{f_1 s_1}{s_1 - f_1} = \frac{(10 \text{ cm})(5 \text{ cm})}{5 \text{ cm} - 10 \text{ cm}} = -10 \text{ cm}$$

or 10 cm behind (to the left of) the mirror. This image now becomes the object for the lens and $s_2 = 20 \text{ cm}$.

$$s_2' = \frac{f_2 s_2}{s_2 - f_2} = \frac{(10 \text{ cm})(20 \text{ cm})}{20 \text{ cm} - 10 \text{ cm}} = 20 \text{ cm}$$

So the image is 20 cm to the right of the lens.

$$m = \left(-\frac{s_1'}{s_1} \right)\left(-\frac{s_2'}{s_2} \right) = \left(-\frac{-10 \text{ cm}}{5 \text{ cm}} \right)\left(-\frac{20 \text{ cm}}{20 \text{ cm}} \right) = -2.0$$

so $h' = hm = (1.0 \text{ cm})(-2.0) = -2.0 \text{ cm}$, where the negative sign indicates the image is inverted.

In summary, both images are 2.0 cm tall; one is upright 10 cm left of the lens, the other is inverted 20 cm to the right of the lens.

(b)

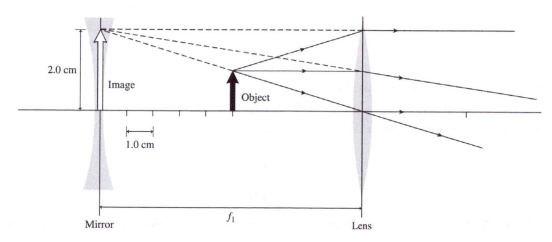

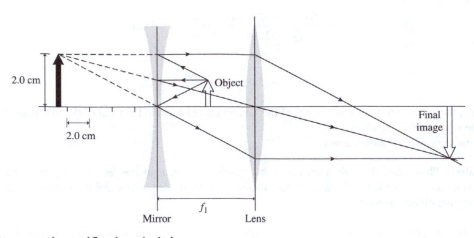

Assess: The ray tracing verifies the calculations.

24.27. Visualize: Hard thought shows that if the left focal points for both lenses coincide then the parallel rays before and after the beam splitter are reproduced. The first lens diverges the rays as if they had come from the focal point of the converging lens.

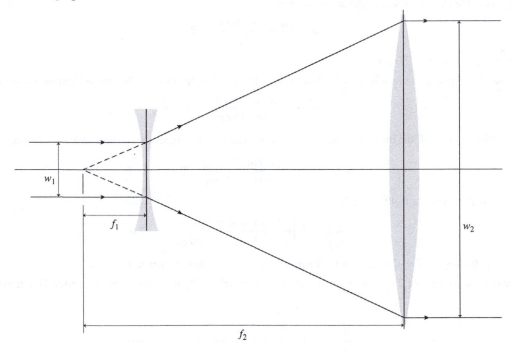

Solve: (a)

$$d = f_2 - |f_1|$$

But since we are given $f_1 < 0$, this is equivalent to

$$d = f_2 + f_1$$

(b) Looking at the similar triangles in the diagram shows that

$$\frac{w_1}{|f_1|} = \frac{w_2}{f_2}$$

$$w_2 = \frac{f_2}{|f_1|} w_1$$

Assess: Figure P24.27 says $f_2 > |f_1|$ and our answer then shows that $w_2 > w_1$ which is the goal of a beam expander.

24.31. Model: Yang has myopia. Normal vision will allow Yang to focus on a very distant object. In measuring distances, we'll ignore the small space between the lens and her eye.

Solve: Because Yang can see objects at 150 cm with a fully relaxed eye, we want a lens that creates a virtual image at $s' = -150$ cm (negative because it's a virtual image) of an object at $s = \infty$ cm. From the thin-lens equation,

$$\frac{1}{f} = \frac{1}{s} + \frac{1}{s'} = \frac{1}{\infty \text{ m}} + \frac{1}{-1.5 \text{ m}} = -0.67 \text{ D}$$

So Yang gets a prescription for a -0.67 D lens which has $f = -150$ cm.

Since Yang can accommodate to see things as close as 20 cm we need to create a virtual image at 20 cm of objects that are at $s = $ new near point. That is, we want to solve the thin-lens equation for s when $s' = -20$ cm and $f = -150$ cm.

$$s = \frac{fs'}{s' - f} = \frac{(-150 \text{ cm})(-20 \text{ cm})}{-20 \text{ cm} - (-150 \text{ cm})} = 23 \text{ cm}$$

Assess: Diverging lenses are always used to correct myopia.

24.33. Visualize: Use Equation 23.27, the lens makers' equation:

$$\frac{1}{f} = (n-1)\left(\frac{1}{R_1} - \frac{1}{R_2}\right)$$

For a symmetric lens $R_1 = R_2$ and

$$f = \frac{R}{2(n-1)} \text{ and } R = 2(n-1)f$$

Also needed will be the magnification of a telescope: $M = -f_{\text{obj}}/f_{\text{eye}} \Rightarrow f_{\text{eye}} = -f_{\text{obj}}/M$ (but we will drop the negative sign).

We are given $R_{\text{obj}} = 100$ cm and $M = 20$.

Solve:

$$R_{\text{eye}} = 2(n-1)f_{\text{eye}} = 2(n-1)\frac{f_{\text{obj}}}{M} = 2(n-1)\frac{\dfrac{R_{\text{obj}}}{2(n-1)}}{M} = \frac{R_{\text{obj}}}{M} = \frac{100 \text{ cm}}{20} = 5.0 \text{ cm}$$

Assess: We expect a short focal length and small radius of curvature for telescope eyepieces.

24.37. Model: While $s \approx f_{\text{obj}}$ we will not assume they are equal.

Visualize: Equation 24.9 says $m_{\text{obj}} \approx -L/f_{\text{obj}}$. We are given $L = 180$ mm and $m_{\text{obj}} = -40$, where the negative sign means the image is inverted.

Solve: Solve for f_{obj}.

$$f_{\text{obj}} = -\frac{L}{m_{\text{obj}}} = \frac{180 \text{ mm}}{40} = 4.5 \text{ mm}$$

From Equation 24.8, $M_{\text{eye}} = (25 \text{ cm})/f_{\text{eye}} \Rightarrow f_{\text{eye}} = 25 \text{ cm}/20 = 1.25 \text{ cm}$. For relaxed eye viewing the image of the objective must be 1.25 cm = 12.5 mm from the eyepiece, so $s' = 180 \text{ mm} - 12.5 \text{ mm} = 167.5 \text{ mm}$. Thus the sample distance is

$$s = \left(\frac{1}{4.5 \text{ mm}} - \frac{1}{167.5 \text{ mm}}\right)^{-1} = 4.6 \text{ mm}$$

Assess: You need a short focal length to achieve 800× magnification. We can also verify that $s \approx f_{\text{obj}}$.

24.41. Model: Two objects are marginally resolved if the angular separation between the objects is $\alpha = 1.22\lambda/D$.

Visualize:

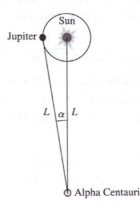

Solve: (a) The angular separation between the sun and Jupiter is

$$\alpha = \frac{780\times10^9 \text{ m}}{4.3 \text{ light years}} = \frac{780\times10^9 \text{ m}}{4.3\times(3.0\times10^8)\times(365\times24\times3600) \text{ m}} = 1.92\times10^{-5} \text{ rad}$$

$$\alpha = \frac{1.22\lambda}{D} = \frac{1.22(600\times10^{-9} \text{ m})}{D} \Rightarrow D = 0.038 \text{ m} = 3.8 \text{ cm}$$

(b) The sun is vastly brighter than Jupiter, which is much smaller and seen only dimly by reflected light. In theory it may be possible to resolve Jupiter and the sun, but in practice the extremely bright light from the sun will overwhelm the very dim light from Jupiter.

ELECTRIC CHARGES AND FORCES

Exercises and Problems

Section 25.1 Developing a Charge Model

Section 25.2 Charge

25.1. Model: Use the charge model.
Solve: (a) In the process of charging by rubbing, electrons are removed from one material and transferred into the other because they are relatively free to move. Protons, on the other hand, are tightly bound in the nuclei of atoms and so are essentially not free to move. Thus, electrons have been added to the plastic rod to make it negatively charged.
(b) Because each electron has a charge of -1.60×10^{-19} C, the number of electrons added is

$$\frac{-12 \times 10^{-9} \text{ C}}{-1.60 \times 10^{-19} \text{ C}} = 7.5 \times 10^{10}$$

25.5. Model: Use the charge model.
Solve: Each helium atom has 2 protons and there are 6.02×10^{23} helium molecules in 1.0 mole of helium. Because each proton has a charge of $+1.60 \times 10^{-19}$ C, the amount of charge in 1.0 mole of oxygen is

$$(1.0 \text{ mol})(6.022 \times 10^{23} \text{ atoms/mol})(2 \text{ protons/atom})(1.6 \times 10^{-19} \text{ C/proton}) = 1.9 \times 10^{5} \text{ C}$$

Section 25.3 Insulators and Conductors

25.9. Model: Use the charge model and the model of a conductor as a material through which electrons move.
Solve:

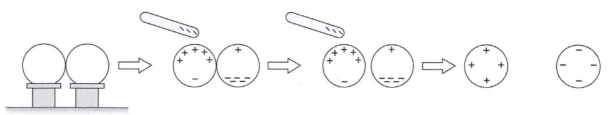

The first step shows two neutral metal spheres touching each other. In the second step, the negative rod repels the negative charges which will retreat as far as possible from the top of the left sphere. Note that the two spheres are touching and the net charge on these two spheres is still zero. While the rod is there on top of the left sphere, the right sphere is moved away from the left sphere. Because the right sphere has an excess negative charge then, by charge conservation, the left sphere has the same magnitude of positive charge. Upon separation, the negative charge is trapped on the right sphere, as shown in the third step. As the two spheres are moved apart farther and the negatively charged rod is moved away from the spheres, the charges on the two spheres redistribute uniformly over the entire surface spheres. Thus, we are left with two oppositely charged spheres.

25.11. Model: Use the charge model and the model of a conductor as a material through which electrons move.
Solve:

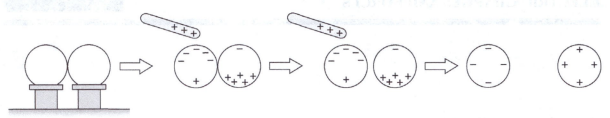

Charging two neutral spheres with opposite charges of equal magnitude can be done through the following four steps. (i) Touch the two neutral metal spheres together. (ii) Bring a charged rod (say, positive) close (but not touching) to one of the spheres (say, the left sphere). Note that the two spheres are still touching and the net charge on the pair is zero. The right sphere has an excess positive charge of exactly the same magnitude as the left sphere's negative charge. (iii) Separate the spheres while the charged rod remains close to the left sphere, so the separated charge remains on the spheres. (iv) Take the charged rod away from the two spheres. The separated charges redistribute uniformly over the metal sphere surfaces.

Section 25.4 Coulomb's Law

25.13. Model: Model the plastic spheres as point charges.
Visualize:

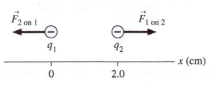

Solve: (a) The charge $q_1 = -50.0$ nC exerts a force $\vec{F}_{1 \text{ on } 2}$ on $q_2 = -50.0$ nC to the right, and the charge q_2 exerts a force $\vec{F}_{2 \text{ on } 1}$ on q_1 to the left. Using Coulomb's law,

$$F_{1 \text{ on } 2} = F_{2 \text{ on } 1} = \frac{K|q_1||q_2|}{r_{12}^2} = \frac{(9.0\times10^9 \text{ N m}^2/\text{C}^2)(50.0\times10^{-9} \text{ C})(50.0\times10^{-9} \text{ C})}{(2.0\times10^{-2} \text{ m})^2} = 0.056 \text{ N}$$

(b) The ratio of the electric force to the weight is

$$\frac{F_{1 \text{ on } 2}}{mg} = \frac{0.056 \text{ N}}{(2.0\times10^{-3} \text{ kg})(9.8 \text{ m/s}^2)} = 2.9$$

25.17. Model: Charges A, B, and C are point charges.
Visualize: Please refer to Figure EX25.17.
Solve: The force on B from charge A is directed downward since two negative charges repel. Coulomb's law gives the magnitude of the force as

$$F_{\text{A on B}} = K\frac{|q_A||q_B|}{r^2} = \frac{(9.0\times10^9 \text{ Nm}^2/\text{C}^2)(1.0\times10^{-9} \text{ C})(2.0\times10^{-9} \text{ C})}{(2.0\times10^{-2} \text{ m })^2} = 4.5\times10^{-5} \text{ N}$$

So $\vec{F}_{\text{A on B}} = (-4.5\times10^{-5} \, \hat{j}) \text{ N}$

The force on B from charge C is directed downwards since the two opposite charges attract. Coulomb's law gives the magnitude of the force as

$$F_{C \text{ on } B} = K \frac{|q_C||q_B|}{r^2} = \frac{(9.0 \times 10^9 \text{ Nm}^2/\text{C}^2)(2.0 \times 10^{-9} \text{ C})(2.0 \times 10^{-9} \text{ C})}{(1.0 \times 10^{-2} \text{ m})^2} = 3.6 \times 10^{-4} \text{ N}$$

So $\vec{F}_{C \text{ on } B} = -(3.6 \times 10^{-4} \, \hat{j}) \text{ N}$

The net electric force on charge A is

$$\vec{F}_A = \vec{F}_{B \text{ on } A} + \vec{F}_{C \text{ on } A} = -(4.5 \times 10^{-5} \, \hat{j}) \text{ N} - (3.6 \times 10^4 \, \hat{j}) \text{ N}$$
$$= -(4.1 \times 10^{-4} \, \hat{j}) \text{ N}$$

25.19. Model: Assume the plastic bead, the proton, and the electron are point charges.
Visualize:

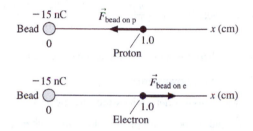

Solve: Coulomb's law gives

$$F_{\text{bead on electron}} = F_{\text{bead on proton}} = \frac{(9.0 \times 10^9 \text{ N m}^2/\text{C}^2)(15 \times 10^{-9} \text{ C})(1.60 \times 10^{-19} \text{ C})}{(1.0 \times 10^{-2} \text{ m})^2} = 2.16 \times 10^{-13} \text{ N}$$

(a) Because the bead is much more massive than both the electron and the proton, we can ignore any acceleration of the bead. Newton's second law is $F = ma$, so

$$a_{\text{proton}} = \frac{F_{\text{bead on proton}}}{m_{\text{proton}}} = \frac{2.16 \times 10^{-13} \text{ N}}{1.67 \times 10^{-27} \text{ kg}} = 1.3 \times 10^{14} \text{ m/s}^2$$

Because opposite charges attract,

$$\vec{a}_{\text{proton}} = (1.3 \times 10^{14} \text{ m/s}^2, \text{ toward bead})$$

(b) Similarly,

$$a_{\text{electron}} = \frac{F_{\text{bead on electron}}}{m_{\text{electron}}} = \frac{2.16 \times 10^{-13} \text{ N}}{9.11 \times 10^{-31} \text{ kg}} = 2.4 \times 10^{17} \text{ m/s}^2$$

Thus $\vec{a}_{\text{electron}} = (2.4 \times 10^{17} \text{ m/s}^2, \text{ away from bead})$.

Section 25.5 The Field Model

25.23. Model: The electric field is that of a negative charge on the plastic bead. Model the bead as a point charge (which is exact for $r > r_{\text{bead}}$).
Solve: The electric field is

$$\vec{E} = K \frac{q}{r^2} \hat{r} = (9.0 \times 10^9 \text{ N m}^2/\text{C}^2) \frac{-8.0 \times 10^{-9} \text{ C}}{(4.0 \times 10^{-2} \text{ m})^2} \hat{r} = -4.5 \times 10^4 \hat{r} \text{ N/C}$$

where $\hat{r}$ is the unit vector from the charge to the point at which we calculate the field. That is, the direction of the electric field is toward the bead.

25.25. Model: A field is the agent that exerts an electric force on a charge.
Visualize:

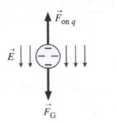

Solve: Newton's second law on the plastic ball is $\Sigma(\vec{F}_{net})_y = F_{on\,q} - F_G$. To balance the gravitational force with the electric force,

$$F_{on\,q} = F_G \implies |q|E = mg \implies E = \frac{mg}{|q|} = \frac{(1.0 \times 10^{-3}\text{ kg})(9.8\text{ N/kg})}{3.0 \times 10^{-9}\text{ C}} = 3.3 \times 10^6\text{ N/C}$$

Because $F_{on\,q}$ must be upward and the charge is negative, the electric field at the location of the plastic ball must be pointing downward. Thus $\vec{E} = (3.3 \times 10^6\text{ N/C, downward})$.

Assess: $\vec{F} = q\vec{E}$ means the sign of the charge q determines the direction of $\vec{F}$ or $\vec{E}$. For positive q, $\vec{E}$ and $\vec{F}$ are pointing in the same direction. But $\vec{E}$ and $\vec{F}$ point in opposite directions when q is negative.

25.29. Model: The beads are point charges.
Visualize:

Plastic bead — 2.0 g — $\vec{F}$ → | ← 2.0 cm → | ← $\vec{F}$ — 4.0 g — Glass bead
−4.0 nC   8.0 nC

Solve: The beads are oppositely charged so are attracted to one another. The force on each is the same by Newton's third law. Coulomb's law give the force as

$$F = K\frac{|q_1||q_2|}{r^2} = (9.0 \times 10^{-9}\text{ Nm}^2/\text{C}^2)\frac{(4.0 \times 10^{-9}\text{ C})(8.0 \times 10^{-9}\text{ C})}{(2.0 \times 10^{-2}\text{ m})^2} = 7.2 \times 10^{-4}\text{ N}$$

The beads accelerate at different rates because their masses are different. By Newton's second law, the acceleration of the plastic bead is

$$a_{plastic} = \frac{F}{m} = \frac{(7.2 \times 10^{-4}\text{ N})}{(2.0 \times 10^{-3}\text{ kg})} = 0.36\text{ m/s}^2 \text{ toward glass bead.}$$

For the glass bead, the acceleration is

$$a_{glass} = \frac{(7.2 \times 10^{-4}\text{ N})}{(4.0 \times 10^{-3}\text{ kg})} = 0.18\text{ m/s}^2 \text{ toward plastic bead.}$$

25.33. Model: The charges are point charges.
Visualize: Please refer to Figure P25.33.
Solve: The electric force on charge q_1 is the vector sum of the forces $\vec{F}_{2\,on\,1}$ and $\vec{F}_{3\,on\,1}$, where q_1 is the 1.0 nC charge, q_2 is the 2.0 nC charge, and q_3 is the other 2.0 nC charge. We have

$$\vec{F}_{2 \text{ on } 1} = \left(\frac{K|q_1||q_2|}{r^2}, \text{ away from } q_2\right)$$

$$= \left(\frac{(9.0\times10^9 \text{ N m}^2/\text{C}^2)(1.0\times10^{-9} \text{ C})(2.0\times10^{-9} \text{ C})}{(1.0\times10^{-2} \text{ m})^2}, \text{ away from } q_2\right)$$

$$= (1.8\times10^{-4} \text{ N, away from } q_2) = (1.8\times10^{-4} \text{ N})[\cos(60°)\hat{i} + \sin(60°)\hat{j}]$$

$$\vec{F}_{3 \text{ on } 1} = \left(\frac{K|q_1||q_3|}{r_2}, \text{ away from } q_3\right) = (1.8\times10^{-4} \text{ N, away from } q_3) = (1.8\times10^{-4} \text{ N})[-\cos(60°)\hat{i} + \sin(60°)\hat{j}]$$

$$\vec{F}_{\text{on } 1} = \vec{F}_{2 \text{ on } 1} + \vec{F}_{3 \text{ on } 1} = 2(1.8\times10^{-4} \text{ N})\sin(60°)\hat{j} = (3.1\times10^{-4}\hat{j}) \text{ N}$$

The force on the 1.0 nC charge is 3.1×10^{-4} N directed upward.

25.35. Model: The charges are point charges.
Visualize:

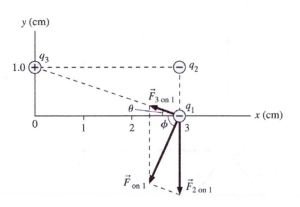

Solve: The electric force on charge q_1 is the vector sum of the forces $\vec{F}_{2 \text{ on } 1}$ and $\vec{F}_{3 \text{ on } 1}$. We have

$$\vec{F}_{2 \text{ on } 1} = \left(\frac{K|q_1||q_2|}{r^2}, \text{ away from } q_2\right)$$

$$= \left(\frac{(9.0\times10^9 \text{ N m}^2/\text{C}^2)(10\times10^{-9} \text{ C})(5.0\times10^{-9} \text{ C})}{(1.0\times10^{-2} \text{ m})^2}, \text{ away from } q_2\right)$$

$$= (4.5\times10^{-3} \text{ N, away from } q_2) = -4.5\times10^{-3}\hat{j} \text{ N}$$

$$\vec{F}_{3 \text{ on } 1} = \left(\frac{K|q_1||q_2|}{r^2}, \text{ toward } q_3\right)$$

$$= \left(\frac{(9.0\times10^9 \text{ N m}^2/\text{C}^2)(10\times10^{-9} \text{ C})(15\times10^{-9} \text{ C})}{(3.0\times10^{-2} \text{ m})^2 + (1.0\times10^{-2} \text{ m})^2}, \text{ toward } q_3\right)$$

$$= (1.35\times10^{-3} \text{ N, toward } q_3) = (1.35\times10^{-3} \text{ N})(-\cos\theta\hat{i} + \sin\theta\hat{j})$$

From the geometry of the figure,

$$\theta = \tan^{-1}\left(\frac{1.0 \text{ cm}}{3.0 \text{ cm}}\right) = 18.4°$$

This means $\cos\theta = 0.949$ and $\sin\theta = 0.316$. Therefore,

$$\vec{F}_{3 \text{ on } 1} = (-1.28\times10^{-3}\hat{i} + 0.43\times10^{-3}\hat{j}) \text{ N}$$

$$\vec{F}_{\text{on } 1} = \vec{F}_{2 \text{ on } 1} + \vec{F}_{3 \text{ on } 1} = (-1.28\times10^{-3}\hat{i} - 4.07\times10^{-3}\hat{j}) \text{ N}$$

The magnitude and direction of the resultant force vector are

$$F_{on\,1} = \sqrt{(-1.28\times10^{-3}\ \text{N})^2 + (-4.07\times10^{-3}\ \text{N})^2} = 4.3\times10^{-3}\ \text{N}$$

$$\tan\phi = \frac{4.07\times10^{-3}\ \text{N}}{1.28\times10^{-3}\ \text{N}} = 3.180 \quad\Rightarrow\quad \phi = \tan^{-1}(3.180) = 73°\ \text{below the }{-x}\text{ axis,}$$

or $F_{on\,1}$ points 253° counterclockwise from the +x-axis.

25.37. Model: The charges are point charges.
Visualize:

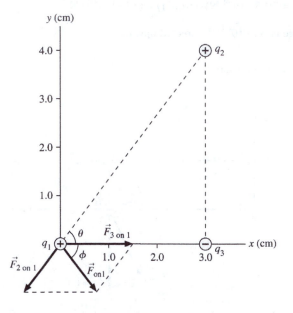

Solve: The electric force on charge q_1 is the vector sum of the forces $\vec{F}_{2\ on\ 1}$ and $\vec{F}_{3\ on\ 1}$. We have

$$\vec{F}_{2\ on\ 1} = \left(\frac{K|q_1||q_2|}{r^2},\ \text{away from } q_2\right)$$

$$= \left(\frac{(9.0\times10^9\ \text{N m}^2/\text{C}^2)(5\times10^{-9}\ \text{C})(10\times10^{-9}\ \text{C})}{(4.0\times10^{-2}\ \text{m})^2 + (3.0\times10^{-2}\ \text{m})^2},\ \text{away from } q_2\right)$$

$$= (1.8\times10^{-4}\ \text{N, away from } q_2) = (1.8\times10^{-4}\ \text{N})(-\cos\theta\,\hat{i} - \sin\theta\,\hat{j})$$

From the geometry of the figure,

$$\tan\theta = \frac{4.0\ \text{cm}}{3.0\ \text{cm}} \quad\Rightarrow\quad \theta = 53.13° \quad\Rightarrow\quad \vec{F}_{2\ on\ 1} = (1.8\times10^{-4}\ \text{N})(-0.60\hat{i} - 0.80\hat{j})$$

$$\vec{F}_{3\ on\ 1} = \left(\frac{(9.0\times10^9\ \text{N m}^2/\text{C}^2)(5\times10^{-9}\ \text{C})(5\times10^{-9}\ \text{C})}{(3.0\times10^{-2}\ \text{m})^2},\ \text{toward } q_3\right) = (2.5\times10^{-4}\ \text{N, toward } q_3) = 2.5\times10^{-4}\,\hat{i}\ \text{N}$$

$$\vec{F}_{on\ 1} = \vec{F}_{2\ on\ 1} + \vec{F}_{3\ on\ 1} = (1.42\times10^{-4}\,\hat{i} - 1.44\times10^{-4}\,\hat{j})\ \text{N}$$

The magnitude and direction of the resultant force vector are

$$F_{on\ 1} = \sqrt{(1.42\times10^{-4}\ \text{N})^2 + (-1.44\times10^{-4}\ \text{N})^2} = 2.0\times10^{-4}\ \text{N}$$

$$\phi = \tan^{-1}\left(\frac{1.44\times10^{-4}\ \text{N}}{1.42\times10^{-4}\ \text{N}}\right) = 45°\ \text{clockwise from the +}x\text{-axis.}$$

25.41. Model: The charges are point charges.

Visualize: Please refer to Figure P25.41.

Solve: Place the 1.0 nC charge at the origin and call it q_1; the −6.0 nC is q_3, the q_2 charge is in the first quadrant, and the q_4 charge is in the second quadrant. The net electric force on q_1 is the vector sum of the electric forces from the other three charges q_2, q_3, and q_4. We have

$$\vec{F}_{2\text{ on }1} = \left(\frac{K|q_1||q_2|}{r^2}, \text{ toward } q_2 \right)$$

$$= \left(\frac{(9.0\times10^9 \text{ N m}^2/\text{C}^2)(1.0\times10^{-9} \text{ C})(2.0\times10^{-9} \text{ C})}{(5.0\times10^{-2} \text{ m})^2}, \text{ toward } q_2 \right)$$

$$= (0.72\times10^{-5} \text{ N, toward } q_2) = (0.72\times10^{-5} \text{ N})[\cos(45°)\hat{i} + \sin(45°)\hat{j}]$$

$$\vec{F}_{3\text{ on }1} = \left(\frac{K|q_1||q_3|}{r^2}, \text{ toward } q_3 \right) = (2.16\times10^{-5} \text{ N, toward } q_3) = 2.16\times10^{-5}\,\hat{j} \text{ N}$$

$$\vec{F}_{4\text{ on }1} = \left(\frac{K|q_1||q_4|}{r^2}, \text{ away from } q_4 \right) = (0.72\times10^{-5} \text{ N})[\cos(45°)\hat{i} - \sin(45°)\hat{j}]$$

Adding these components together vector-wise gives

$$\vec{F}_{\text{on }1} = \vec{F}_{2\text{ on }1} + \vec{F}_{3\text{ on }1} + \vec{F}_{4\text{ on }1} = 2(0.72\times10^{-5} \text{ N})\cos(45°)\hat{i} + (2.16\times10^{-5} \text{ N})\hat{j}$$

$$= (1.02\times10^{-5}\,\hat{i} + 2.2\times10^{-5}\,\hat{j}) \text{ N}$$

25.45. Model: The charged particles are point charges.

Visualize:

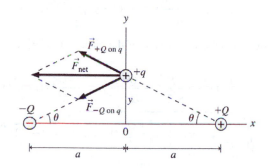

Solve: The force on q is the vector sum of the force from $-Q$ and $+Q$. We have

$$\vec{F}_{-Q\text{ on }+q} = \left(K\frac{|-Q||+q|}{a^2 + y^2}, \text{ toward } -Q \right) = \frac{KQq}{a^2 + y^2}(-\cos\theta\,\hat{i} - \sin\theta\,\hat{j})$$

$$\vec{F}_{+Q\text{ on }+q} = \left(\frac{K|+Q||+q|}{a^2 + y^2}, \text{ away from } +Q \right) = \frac{KQq}{a^2 + y^2}(-\cos\theta\,\hat{i} + \sin\theta\,\hat{j})$$

$$\vec{F}_{\text{net}} = \frac{KQq}{a^2 + y^2}(-2\cos\theta)\hat{i} + 0\,\hat{j} \text{ N}$$

From the figure we see that $\cos\theta = a/\sqrt{a^2 + y^2}$. Thus

$$(F_{\text{net}})_x = \frac{-2KQqa}{(a^2 + y^2)^{3/2}}$$

Assess: Note that $(F_{\text{net}})_x = F_{\text{net}}$ because the y-components of the two forces cancel each other out.

25.47. Model: The charges are point charges.
Solve:

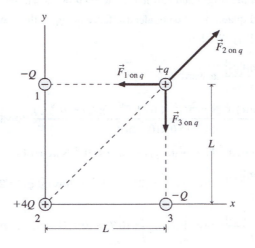

We will denote the charges $-Q$, $4Q$ and $-Q$ by 1, 2, and 3, respectively.

$$\vec{F}_{1 \text{ on } q} = \left(\frac{K|-Q||q|}{L^2}, \text{ toward } -Q \right) = \frac{KQq}{L^2}(-\hat{i})$$

$$\vec{F}_{2 \text{ on } q} = \left(\frac{K|4Q||q|}{(\sqrt{2}L)^2}, \text{ away from } 4Q \right) = \frac{4KQq}{2L^2}[+\cos(45°)\hat{i} + \sin(45°)\hat{j}] \Rightarrow F_{2 \text{ on } q} = \frac{2KQq}{L^2} = 2F_{1 \text{ on } q}$$

$$\vec{F}_{3 \text{ on } q} = \left(\frac{K|-Q||q|}{L^2}, \text{ toward } -Q \right) = \frac{KQq}{L^2}(-\hat{j})$$

The net electric force on the charge $+q$ is the vector sum of the electric forces from the other three charges. The net force is

$$\vec{F}_{\text{net}} = \frac{KQq}{L^2}(-\hat{i}) + \frac{2KQq}{L^2}\left(+\frac{\hat{i}}{\sqrt{2}} + \frac{\hat{j}}{\sqrt{2}} \right) + \frac{KQq}{L^2}(-\hat{j}) = -\frac{KQq}{L^2}\hat{i}(1-\sqrt{2}) - \frac{KQq}{L^2}\hat{j}(1-\sqrt{2})$$

$$F_{\text{net}} = \sqrt{\left[\frac{KQq}{L^2}(1-\sqrt{2}) \right]^2 + \left[\frac{KQq}{L^2}(1-\sqrt{2}) \right]^2} = (2-\sqrt{2})\frac{KQq}{L^2}$$

25.49. Model: The charges are point charges.
Visualize:

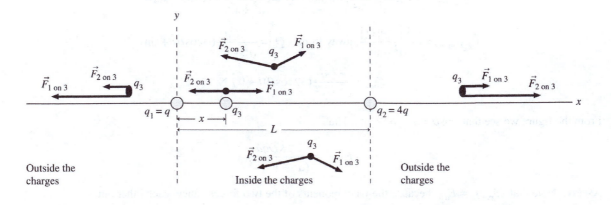

We must first identify the region of space where the third charge q_3 is located. You can see from the figure that the forces can't possibly add to zero if q_3 is above or below the axis or outside the charges. However, at some point on the x-axis between the two charges the forces from the two charges will be oppositely directed.

Solve: The mathematical problem is to find the position for which the forces $\vec{F}_{1 \text{ on } 3}$ and $\vec{F}_{2 \text{ on } 3}$ are equal in magnitude. If q_3 is the distance x from q_1, it is the distance $L - x$ from q_2. The magnitudes of the forces are

$$F_{1 \text{ on } 3} = \frac{K|q_1||q_3|}{r_{13}^2} = \frac{Kq|q_3|}{x^2} \qquad F_{2 \text{ on } 3} = \frac{K|q_2||q_3|}{r_{23}^2} = \frac{K(4q)|q_3|}{(L-x)^2}$$

Equating the two forces gives

$$\frac{Kq|q_3|}{x^2} = \frac{K(4q)|q_3|}{(L-x)^2} \quad \Rightarrow \quad (L-x)^2 = 4x^2 \quad \Rightarrow \quad x = \frac{L}{3} \text{ and } -L$$

The solution $x = -L$ is not allowed as you can see from the figure. To find the magnitude of the charge q_3, we apply the equilibrium condition to charge q_1:

$$F_{2 \text{ on } 1} = F_{3 \text{ on } 1} \quad \Rightarrow \quad \frac{K|q_2||q_1|}{L^2} = \frac{K|q_3||q_1|}{\left(\frac{1}{3}L\right)^2} \quad \Rightarrow \quad 4q = 9|q_3| \quad \Rightarrow \quad |q_3| = \frac{4}{9}q$$

We are now able to check the static equilibrium condition for the charge $4q$ (or q_2):

$$F_{1 \text{ on } 2} = F_{3 \text{ on } 2} \quad \Rightarrow \quad K\frac{|q_1||q_2|}{L^2} = \frac{K|q_3||q_2|}{(L-x)^2} \quad \Rightarrow \quad \frac{q}{L^2} = \frac{\frac{4}{9}q}{\left(\frac{2}{3}L\right)^2} = \frac{q}{L^2}$$

The sign of the third charge q_3 must be negative. A positive sign on q_3 will not have a net force of zero either on the charge q or the charge $4q$. In summary, a charge of $-\frac{4}{9}q$ placed $x = \frac{1}{3}L$ from the charge q will cause the 3-charge system to be in static equilibrium.

25.53. Model: Model the bee as a point charge.
Solve: (a) The force on the bee due to gravity is $F_G = mg$, and the electric force on the bee is $F_e = Eq$. The ratio of the electric force to the bee's weight is

$$\frac{F_e}{F_G} = \frac{Eq}{mg} = \frac{(100 \text{ N/C})(23 \times 10^{-12} \text{ C})}{(0.10 \times 10^{-3} \text{ kg})(9.8 \text{ m/s}^2)} = 2.3 \times 10^{-6}$$

(b) For the bee to be suspended by the electric force, this force must have the same magnitude as the force due to gravity and be directed upward. Equating the magnitudes of the forces and solving for E give

$$Eq = mg \quad \Rightarrow \quad E = \frac{mg}{q} = \frac{(0.10 \times 10^{-3} \text{ kg})(9.8 \text{ m/s}^2)}{(23 \times 10^{-12} \text{ C})} = 4.3 \times 10^7 \text{ N/C}$$

The electric field must be directed upward so that the force on a positive charge is upward.

25.55. Model: The charged plastic beads are point charges and the spring is an ideal spring that obeys Hooke's law.
Solve: Let q be the charge on each plastic bead. The repulsive force between the beads pushes the beads apart. The spring is stretched until the restoring spring force on either bead is equal to the repulsive Coulomb force:

$$\frac{Kq^2}{r^2} = k\Delta x \quad \Rightarrow \quad q = \sqrt{\frac{k\Delta x\, r^2}{K}}$$

The spring constant k is obtained by noting that the weight of a 1.0 g mass stretches the spring 1.0 cm. Thus

$$mg = k(1.0 \times 10^{-2} \text{ m}) \quad \Rightarrow \quad k = \frac{(1.0 \times 10^{-3} \text{ kg})(9.8 \text{ N/kg})}{1.0 \times 10^{-2} \text{ m}} = 0.98 \text{ N/m}$$

$$q = \sqrt{\frac{(0.98 \text{ N/m})(4.5 \times 10^{-2} \text{ m} - 4.0 \times 10^{-2} \text{ m})(4.5 \times 10^{-2} \text{ m})^2}{9.0 \times 10^9 \text{ N m}^2/\text{C}^2}} = 33 \text{ nC}$$

25.57. Solve: (a) Kinetic energy is $K = \frac{1}{2}mv^2$, so the velocity squared is $v^2 = 2K/m$. From kinematics, a particle moving through distance Δx with acceleration a, starting from rest, finishes with $v^2 = 2a\Delta x$. To gain $K = 2 \times 10^{-18}$ J of kinetic energy in $\Delta x = 2.0$ μm requires an acceleration of

$$a = \frac{v^2}{2\Delta x} = \frac{2K/m}{2\Delta x} = \frac{K}{m\Delta x} = \frac{2.0 \times 10^{-18} \text{ J}}{(9.11 \times 10^{-31} \text{ kg})(2.0 \times 10^{-6} \text{ m})} = 1.10 \times 10^{18} \text{ m/s}^2 \approx 1.1 \times 10^{18} \text{ m/s}^2$$

(b) The force that produces this acceleration is

$$F = ma = (9.11 \times 10^{-31} \text{ kg})(1.10 \times 10^{18} \text{ m/s}^2) = 1.00 \times 10^{-12} \text{ N} \approx 1.0 \times 10^{-12} \text{ N}$$

(c) The electric field required is

$$E = \frac{F}{e} = \frac{1.00 \times 10^{-12} \text{ N}}{1.602 \times 10^{-19} \text{ C}} = 6.3 \times 10^6 \text{ N/C}$$

(d) The force on an electron due to a charge q is $F = K|q|e/r^2$. To have breakdown, the force on the electron must be at least 1.0×10^{-12} N. The *minimum* charge that could cause a breakdown will be the charge that causes exactly a force of 1.0×10^{-12} N:

$$|\vec{F}| = \frac{K|q|e}{r^2} = 1.0 \times 10^{-12} \text{ N} \implies |q| = \frac{r^2 F}{Ke} = \frac{(0.010 \text{ m})^2 (1.0 \times 10^{-12} \text{ N})}{(9.0 \times 10^9 \text{ N m}^2/\text{C}^2)(1.6 \times 10^{-19} \text{ C})} = 6.9 \times 10^{-8} \text{ C} = 69 \text{ nC}$$

25.61. Model: The electric field is that of a negative point charge.
Visualize: Please refer to Figure P25.61. Place point 1 at the origin.
Solve: The electric field is

$$\vec{E} = \left(K\frac{|q|}{r^2}, \text{ toward } q \right) = \left(\frac{(9.0 \times 10^9 \text{ N m}^2/\text{C}^2)(2.0 \times 10^{-9} \text{ C})}{r^2}, \text{ toward } q \right)$$

$$= \left(\frac{18.0 \text{ N m}^2/\text{C}}{r^2}, \text{ toward } q \right)$$

The electric fields at the two points are

$$\vec{E}_1 = \left(\frac{18.0 \text{ N m}^2/\text{C}}{(1.0 \times 10^{-2} \text{ m})^2}, \text{ toward } q \right)$$

$$= (1.8 \times 10^5 \text{ N/C}, 60° \text{ counter clockwise from the } +x\text{-axis or } 60° \text{ north of east})$$

$$\vec{E}_2 = \left(\frac{18.0 \text{ N m}^2/\text{C}}{(1.0 \times 10^{-2} \text{ m})^2}, \text{ toward } q \right)$$

$$= (1.8 \times 10^5 \text{ N/C}, 60° \text{ clockwise from the } -x\text{-axis or } 60° \text{ north of west})$$

25.65. Model: The electric field is that of three point charges.
Visualize:

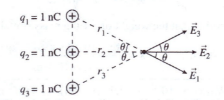

Solve: **(a)** In the figure, the distances are $r_1 = r_3 = \sqrt{(1.0 \text{ cm})^2 + (3.0 \text{ cm})^2} = 3.162$ cm and the angle is $\theta = \tan^{-1}(1.0/3.0) = 18.43°$. Using the equation for the field of a point charge,

$$E_1 = E_3 = \frac{K|q_1|}{r_1^2} = \frac{(9.0 \times 10^9 \text{ N m}^2/\text{C}^2)(1.0 \times 10^{-9} \text{ C})}{(0.03162 \text{ m})^2} = 9.0 \text{ kN/C}$$

We now use the angle θ to find the components of the field vectors:

$$\vec{E}_1 = E_1 \cos\theta \, \hat{i} - E_1 \sin\theta \, \hat{j} = (8540\hat{i} - 2840\,\hat{j}) \text{ N/C} = (8.5\hat{i} - 2.8\,\hat{j}) \text{ kN/C}$$
$$\vec{E}_3 = E_3 \cos\theta \, \hat{i} + E_3 \sin\theta \, \hat{j} = (8540\hat{i} + 2840\,\hat{j}) \text{ N/C} = (8.5\hat{i} + 2.8\,\hat{j}) \text{ kN/C}$$

$\vec{E}_2$ is easier since it has only an x-component. Its magnitude is

$$E_2 = \frac{K|q_2|}{r_2^2} = \frac{(9.0 \times 10^9 \text{ N m}^2/\text{C}^2)(1.0 \times 10^{-9} \text{ C})}{(0.0300 \text{ m})^2} = 10{,}000 \text{ N/C} \quad \Rightarrow \quad \vec{E}_2 = E_2 \hat{i} = 10\hat{i} \text{ kN/C}$$

(b) The electric field is defined in terms of an electric *force* acting on charge q: $\vec{E} = \vec{F}/q$. Since forces obey a principle of superposition ($\vec{F}_{\text{net}} = \vec{F}_1 + \vec{F}_2 + \cdots$) it follows that the electric field due to several charges also obeys a principle of superposition.

(c) The net electric field at a point 3.0 cm to the right of q_2 is $\vec{E}_{\text{net}} = \vec{E}_1 + \vec{E}_2 + \vec{E}_3 = 27\hat{i}$ kN/C. The y-components of $\vec{E}_1$ and $\vec{E}_2$ cancel, giving a net field pointing along the x-axis.

THE ELECTRIC FIELD

Exercises and Problems

Section 26.2 The Electric Field of Multiple Point Charges

26.3. Model: The electric field is that due to superposition of the fields of the two 3.0 nC charges located on the y-axis.
Visualize: We denote the top 3.0 nC charge by q_1 and the bottom 3.0 nC charge by q_2. The electric fields ($\vec{E}_1$ and $\vec{E}_2$) of both the positive charges are directed away from their respective charges. With vector addition, they yield the net electric field $\vec{E}_{net}$ at the point P indicated by the dot.
Solve: The electric fields from q_1 and q_2 are

$$\vec{E}_1 = \left(\frac{1}{4\pi\varepsilon_0} \frac{|q_1|}{r_1^2}, \text{ along } +x\text{-axis} \right) = \frac{(9.0\times10^9 \text{ N m}^2/\text{C}^2)(3.0\times10^{-9} \text{ C})}{(0.05 \text{ m})^2}\hat{i} = 10{,}800\hat{i} \text{ N/C}$$

$$\vec{E}_2 = \left(\frac{1}{4\pi\varepsilon_0} \frac{|q_2|}{r_2^2}, \theta \text{ below } -y\text{-axis} \right)$$

Because $\tan\theta = 10 \text{ cm}/5 \text{ cm}$, $\theta = \tan^{-1}(2) = 63.43°$. So,

$$\vec{E}_2 = \frac{(9.0\times10^9 \text{ N m}^2/\text{C}^2)(3.0\times10^{-9} \text{ C})}{(0.10 \text{ m})^2 + (0.050 \text{ m})^2}(-\cos 63.43° \, \hat{i} - \sin 63.43° \, \hat{j}) = (-966\hat{i} - 1932\hat{j}) \text{ N/C}$$

The net electric field is thus

$$\vec{E}_{\text{net at P}} = \vec{E}_1 + \vec{E}_2 = (9834\hat{i} - 1932\hat{j}) \text{ N/C}$$

To find the angle this net vector makes with the x-axis, we calculate

$$\tan\phi = \frac{-1932 \text{ N/C}}{9834 \text{ N/C}} \Rightarrow \phi = -11.11°$$

Thus, the strength of the electric field at P is

$$E_{\text{net}} = \sqrt{(9834 \text{ N/C})^2 + (1932 \text{ N/C})^2} = 10021 \text{ N/C} = 10000 \text{ N/C}$$

and $\vec{E}_{net}$ makes an angle of 11.1° below the $+x$-axis.
Assess: Because of the inverse square dependence on distance, $E_2 < E_1$. Additionally, because the point P has no special symmetry relative to the charges, we expected the net field to be at an angle relative to the x-axis.

26.5. Model: The distances to the observation points are large compared to the size of the dipole, so model the field as that of a dipole moment.
Visualize:

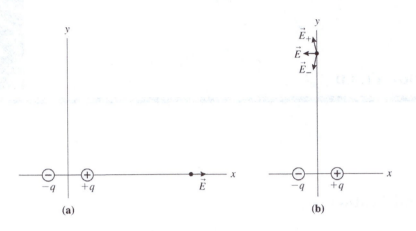

(a)   (b)

The dipole consists of charges $\pm q$ along the x-axis. The electric field in (a) points right. The field in (b) points left.
Solve: **(a)** The dipole moment is

$$\vec{p} = (qs, \text{ from} - \text{to} +) = (1.0 \times 10^{-9} \text{ C})(0.0020 \text{ m})\hat{i} = 2.0 \times 10^{-12}\hat{i} \text{ C m}$$

The electric field at (10 cm, 0 cm), which is at distance $r = 0.10$ m along the axis of the dipole, is

$$\vec{E} = \frac{1}{4\pi\varepsilon_0}\frac{2\vec{p}}{r^3} = (9.0 \times 10^9 \text{ N m}^2/\text{C}^2)\frac{2(2.0 \times 10^{-12}\hat{i} \text{ C m})}{(0.10 \text{ m})^3} = 36\hat{i} \text{ N/C}$$

The field strength, which is all we're asked for, is 36 N/C.
(b) The electric field at (0 cm, 10 cm), which is at $r = 0.10$ m in the plane perpendicular to the electric dipole, is

$$\vec{E} = -\frac{1}{4\pi\varepsilon_0}\frac{\vec{p}}{r^3} = -(9.0 \times 10^9 \text{ N m}^2/\text{C}^2)\frac{2.0 \times 10^{-12}\hat{i} \text{ C m}}{(0.10 \text{ m})^3} = -18.0\hat{i} \text{ N/C}$$

The field strength at this point is 18 N/C.

Section 26.3 The Electric Field of a Continuous Charge Distribution

26.7. Model: We will assume that the wire is thin and that the charge lies on the wire along a *line*.
Solve: From Equation 26.15, the electric field strength of an infinitely long line of charge having linear charge density λ is

$$E_{\text{line}} = \frac{1}{4\pi\varepsilon_0}\frac{2|\lambda|}{r}$$

$$\Rightarrow E_{\text{line}}(r = 5.0 \text{ cm}) = \frac{1}{4\pi\varepsilon_0}\frac{2|\lambda|}{5.0 \times 10^{-2} \text{ m}} \qquad E_{\text{line}}(r = 10.0 \text{ cm}) = \frac{1}{4\pi\varepsilon_0}\frac{2|\lambda|}{10.0 \times 10^{-2} \text{ m}}$$

Dividing the above two equations gives

$$E_{\text{line}}(r = 5.0 \text{ cm}) = \left(\frac{10.0 \times 10^{-2} \text{ m}}{5.0 \times 10^{-2} \text{ m}}\right)E_{\text{line}}(r = 10.0 \text{ cm}) = 2(2000 \text{ N/C}) = 4.00 \times 10^3 \text{ N/C} = 4000 \text{ N/C}$$

Section 26.4 The Electric Field of Rings, Disks, Planes, and Spheres

26.11. Model: Assume that the rings are thin and that the charge lies along circle of radius R.

Visualize:

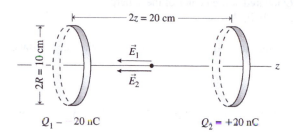

The rings are centered on the z-axis.

Solve: (a) According to Example 26.4, the field of the left (negative) ring at $z = 10$ cm is

$$(E_1)_z = \frac{zQ_1}{4\pi\varepsilon_0(z^2 + R^2)^{3/2}} = \frac{(9.0\times10^9 \text{ N m}^2/\text{C}^2)(0.10 \text{ m})(-20\times10^{-9} \text{ C})}{[(0.10 \text{ m})^2 + (0.050 \text{ m})^2]^{3/2}} = -1.288\times10^4 \text{ N/C}$$

That is, the field is $\vec{E}_1 = (1.288\times10^4 \text{ N/C, left})$. Ring 2 has the same quantity of charge and is at the same distance, so it will produce a field of the same strength. Because Q_2 is positive, $\vec{E}_2$ will also point to the left. The net field at the midpoint is

$$\vec{E} = \vec{E}_1 + \vec{E}_2 = (2.6\times10^4 \text{ N/C, left})$$

(b) The force is

$$\vec{F} = q\vec{E} = (-1.0\times10^{-9} \text{ C})(2.6\times10^4 \text{ N/C, left}) = (2.6\times10^{-5} \text{ N, right})$$

26.13. Model: Each disk is a uniformly charged disk. When the disk is charged negatively, the on-axis electric field of the disk points toward the disk. The electric field points away from the disk for a positively charged disk.
Visualize:

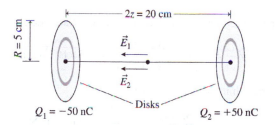

Solve: (a) The surface charge density on the disk is

$$|\eta| = \frac{|Q|}{A} = \frac{|Q|}{\pi R^2} = \frac{50\times10^{-9} \text{ C}}{\pi(0.050 \text{ m})^2} = 6.366\times10^{-6} \text{ C/m}^2$$

From Equation 26.23, the electric field of the left disk at $z = 0.10$ m is

$$(E_1)_z = \frac{\eta}{2\varepsilon_0}\left[1 - \frac{1}{\sqrt{1 + R^2/z^2}}\right] = \frac{-6.366\times10^{-6} \text{ C/m}^2}{2(8.85\times10^{-12} \text{ C}^2/\text{N m}^2)}\left[1 - \frac{1}{\sqrt{1 + (0.050 \text{ m}/0.10 \text{ m})^2}}\right] = -38,000 \text{ N/C}$$

In other words, $\vec{E}_1 = (38,000 \text{ N/C, left})$. Similarly, the electric field of the right disk at $z = 0.10$ m (to its left) is $\vec{E}_2 = (38,000 \text{ N/C, left})$. The net field at the midpoint between the two rings is $\vec{E} = \vec{E}_1 + \vec{E}_2 = (7.6\times10^4 \text{ N/C, left})$.
(b) The force on the charge is

$$\vec{F} = q\vec{E} = (-1.0\times10^{-9} \text{ C})(7.6\times10^4 \text{ N/C, left}) = (7.6\times10^{-5} \text{ N, right})$$

Assess: Note that the force on the negative charge is to the right because the electric field is to the left.

26.15. Model: A spherical shell of charge Q and radius R has an electric field *outside* the sphere that is exactly the same as that of a point charge Q located at the center of the sphere.
Visualize: In the case of a metal ball, the charge resides on its surface. This can then be visualized as a charged spherical shell of radius R.
Solve: Equation 26.28 gives the electric field of a charged spherical shell at a distance $r > R$:

$$\vec{E}_{\text{ball}} = \frac{Q}{4\pi\varepsilon_0 r^2}\hat{r}$$

In the present case, $E_{\text{ball}} = 50{,}000$ N/C at $r = \frac{1}{2}(10 \text{ cm}) + 2.0 \text{ cm} = 7.0 \text{ cm} = 0.070 \text{ m}$. So,

$$Q = 4\pi\varepsilon_0 r^2 E_{\text{ball}} = \frac{(0.070 \text{ m})^2 \, 50{,}000 \text{ N/C}}{9.0\times10^9 \text{ N m}^2/\text{C}^2} = 2.7\times10^{-8} \text{ C} = 27 \text{ nC}$$

Section 26.6 Motion of a Charged Particle in an Electric Field

26.21. Model: The disks form a parallel-plate capacitor. The electric field inside a parallel-plate capacitor is a uniform field, so the proton will have a constant acceleration.
Visualize:

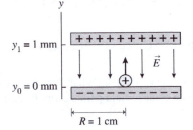

Solve: (a) The two disks form a parallel-plate capacitor with surface charge density

$$\eta = \frac{Q}{A} = \frac{Q}{\pi R^2} = \frac{10\times10^{-9} \text{ C}}{\pi (0.010 \text{ m})^2} = 3.18\times10^{-5} \text{ C/m}^2$$

From Equation 26.29, the field strength inside a capacitor is

$$E = \frac{\eta}{\varepsilon_0} = \frac{3.18\times10^{-5} \text{ C/m}^2}{8.85\times10^{-12} \text{ C}^2/\text{N m}^2} = 3.6\times10^6 \text{ N/C}$$

(b) The electric field points toward the negative plate, so in the coordinate system of the figure $\vec{E} = -3.6\times10^6\,\hat{j}$ N/C. The field exerts forces $\vec{F} = q_{\text{proton}}\vec{E} = e\vec{E}$ on the proton, causing an acceleration with a y-component that is

$$a_y = \frac{F}{m} = \frac{eE_y}{m} = \frac{(1.6\times10^{-19} \text{ C})(-3.6\times10^6 \text{ N/C})}{1.67\times10^{-27} \text{ kg}} = -3.45\times10^{14} \text{ m/s}^2$$

After the proton is launched, this acceleration will cause it to lose speed. To just barely reach the positive plate, it should reach $v_1 = 0$ m/s at $y_1 = 1$ mm. The kinematic equation of motion is

$$v_1^2 = 0 \text{ m}^2/\text{s}^2 = v_0^2 + 2a_y\Delta y$$

$$\Rightarrow v_0 = \sqrt{-2a_y\Delta y} = \sqrt{-2(-3.45\times10^{14} \text{ m/s}^2)(0.0010 \text{ m})} = 8.3\times10^5 \text{ m/s}$$

Assess: The acceleration of the proton in the electric field is enormous in comparison to the gravitation acceleration g. That is why we did not explicitly consider g in our calculations.

26.23. Model: The infinite negatively charged plane produces a uniform electric field that is directed toward the plane.

Visualize:

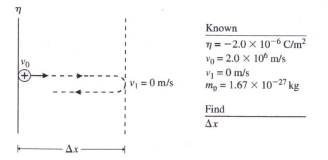

Known

$\eta = -2.0 \times 10^{-6}$ C/m^2
$v_0 = 2.0 \times 10^6$ m/s
$v_1 = 0$ m/s
$m_p = 1.67 \times 10^{-27}$ kg

Find

Δx

Solve: From the kinematic equation of motion $v_1^2 = 0 = v_0^2 + 2a\Delta x$ and $F = qE = ma$,

$$a = \frac{qE}{m} = \frac{-v_0^2}{2\Delta x} \Rightarrow \Delta x = \frac{-mv_0^2}{2qE}$$

Furthermore, the electric field of a plane of charge with surface charge density η is $E = \eta/2\varepsilon_0$. Thus,

$$\Delta x = \frac{-mv_0^2 \varepsilon_0}{q\eta} = \frac{-(1.67 \times 10^{-27} \text{ kg})(2.0 \times 10^6 \text{ m/s})^2 (8.85 \times 10^{-12} \text{ C}^2/\text{N m}^2)}{(1.60 \times 10^{-19} \text{ C})(-2.0 \times 10^{-6} \text{ C/m}^2)} = 0.18 \text{ m}$$

Section 26.7 Motion of a Dipole in an Electric Field

26.25. Model: The external electric field exerts a torque on the dipole moment of the water molecule.
Solve: From Equation 26.34, the torque exerted on a dipole moment by an electric field is $\tau = pE\sin\theta$. The maximum torque is exerted when $\sin\theta = 1$ or $\theta = 90°$. Thus,

$$\tau_{max} = pE = (6.2 \times 10^{-30} \text{ C m})(5.0 \times 10^8 \text{ N/C}) = 3.1 \times 10^{-21} \text{ N m}$$

26.29. Model: The electric field is that of three point charges q_1, q_2, and q_3. Assume the charges are in the *x-y* plane.
Visualize: The -5.0 nC charge is q_1, the bottom 10 nC charge is q_2, and the top 10 nC charge is q_3. The net electric field at the dot is $\vec{E}_{net} = \vec{E}_1 + \vec{E}_2 + \vec{E}_3$. The procedure will be to find the magnitudes of the electric fields, to write them in component form, and to add the components.
Solve: (a) The electric field produced by q_1 is

$$E_1 = \frac{1}{4\pi\varepsilon_0} \frac{|q_1|}{r_1^2} = \frac{(9.0 \times 10^9 \text{ N m}^2/\text{C}^2)(5.0 \times 10^{-9} \text{ C})}{(0.020 \text{ m})^2} = 112,500 \text{ N/C}$$

$\vec{E}_1$ points toward q_1, so in component form $\vec{E}_1 = 112,500\,\hat{j}$ N/C.
The electric field produced by q_2 is $E_2 = 56,250$ N/C. $\vec{E}_2$ points away from q_2, so $\vec{E}_2 = -56,250\,\hat{i}$ N/C.
Finally, the electric field produced by q_3 is

$$E_3 = \frac{1}{4\pi\varepsilon_0} \frac{|q_3|}{r_3^2} = \frac{(9.0 \times 10^9 \text{ N m}^2/\text{C}^2)(10 \times 10^{-9} \text{ C})}{(0.020 \text{ m})^2 + (0.040 \text{ m})^2} = 45,000 \text{ N/C}$$

$\vec{E}_3$ points away from q_3 and makes an angle $\phi = \tan^{-1}(2/4) = 26.6°$ with the *x*-axis. So,

$$\vec{E}_3 = -E_3\cos\phi\,\hat{i} - E_3\sin\phi\,\hat{j} = (-40,250\,\hat{i} - 20,130\,\hat{j}) \text{ N/C}$$

Adding these three vectors gives

$$\vec{E}_{net} = \vec{E}_1 + \vec{E}_2 + \vec{E}_3 = (-96,500\hat{i} + 92,400\hat{j}) \text{ N/C} = (-9.7\times10^4\hat{i} + 9.2\times10^4\hat{j}) \text{ N/C}$$

This is in component form.
(b) The magnitude of the field is

$$E_{net} = \sqrt{E_x^2 + E_y^2} = \sqrt{(96,500 \text{ N/C})^2 + (92,400 \text{ N/C})^2} = 133,600 \text{ N/C} = 1.34\times10^5 \text{ N/C}$$

and its angle below the $-x$-axis is $\theta = \tan^{-1}(|E_y/E_x|) = 44°$. We can also write $\vec{E}_{net} = (1.34\times10^5 \text{ N/C}, 136° \text{ CCW}$ from the $+x$-axis).

26.33. Model: The electric field is that of two positive charges.
Visualize:

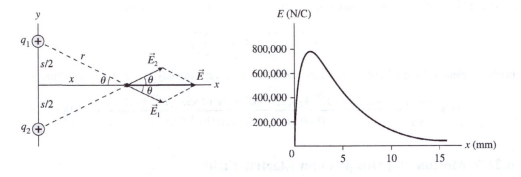

The figure shows $\vec{E}_1$ and $\vec{E}_2$ due to the individual charges. The total field is $\vec{E} = \vec{E}_1 + \vec{E}_2$.
Solve: (a) From symmetry, the y-components of the two electric fields cancel out. The x-components are equal and add. Thus,

$$\vec{E} = 2(E_1)_x\hat{i} = 2E_1\cos\theta\hat{i}$$

The field strength and angle due to q_1 are

$$E_1 = \frac{|q|}{4\pi\varepsilon_0 r^2} = \frac{|q|}{4\pi\varepsilon_0(x^2+s^2/4)} \qquad \cos\theta = \frac{x}{r} = \frac{x}{\sqrt{x^2+s^2/4}}$$

Thus the magnitude of the net field is $E = \dfrac{2qx}{4\pi\varepsilon_0(x^2+s^2/4)^{3/2}}$.

(b) The electric field strength at position x is

$$E = \frac{(9.0\times10^9 \text{ N m}^2/\text{C}^2)(2)(1.0\times10^{-9} \text{ C})x}{[x^2+(0.003 \text{ m})^2]^{3/2}} = \frac{18x}{[x^2+(0.003 \text{ m})^2]^{3/2}} \text{ N/C}$$

where x has to be in meters. We can now evaluate E for different values of x:

x (mm)	x (m)	E (N/C)
0	0.000	0
2	0.002	768,000
4	0.004	576,000
6	0.006	358,000
10	0.010	158,000

26.35. Model: The electric field is that of three point charges.
Visualize:

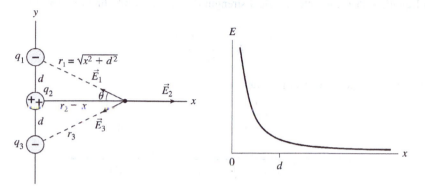

The field at points on the x-axis is $\vec{E}_{net} = \vec{E}_1 + \vec{E}_2 + \vec{E}_3$.

Solve: (a) We note from the symmetry that the y-component of $\vec{E}_1$ and $\vec{E}_3$ cancel. Since $\vec{E}_2$ has no y-component, the net field will have only an x-component. The x-components of $\vec{E}_1$ and $\vec{E}_3$ are equal, so

$$(E_{net})_x = E_2 - 2E_1 \cos\theta \qquad (E_{net})_y = (E_{net})_z = 0 \text{ N/C}$$

Note that the signs of q_1 and q_3 were used in writing this equation. The electric field strength due to q_2 is

$$E_2 = \frac{1}{4\pi\varepsilon_0} \frac{|q_2|}{r_2^2} = \frac{1}{4\pi\varepsilon_0} \frac{2q}{x^2}$$

The electric field strength due to q_1 is

$$E_1 = \frac{1}{4\pi\varepsilon_0} \frac{|q_1|}{r_1^2} = \frac{1}{4\pi\varepsilon_0} \frac{q}{x^2 + d^2}$$

From geometry,

$$\cos\theta = \frac{x}{r_1} = \frac{x}{\sqrt{x^2 + d^2}}$$

Assembling these pieces, the net field is

$$(E_{net})_x = \frac{1}{4\pi\varepsilon_0} \frac{2q}{x^2} - \frac{1}{4\pi\varepsilon_0} \frac{2q}{x^2 + d^2} \frac{x}{\sqrt{x^2 + d^2}} = \frac{2q}{4\pi\varepsilon_0}\left[\frac{1}{x^2} - \frac{x}{(x^2 + d^2)^{3/2}}\right]$$

$$\Rightarrow \vec{E}_{net} = \frac{2q}{4\pi\varepsilon_0}\left[\frac{1}{x^2} - \frac{x}{(x^2 + d^2)^{3/2}}\right]\hat{i}$$

(b) For $x \ll d$, the observation point is very close to $q_2 = +2q$. Furthermore, at $x \approx 0$ m the fields $\vec{E}_1$ and $\vec{E}_3$ are nearly opposite to each other and will nearly cancel. So for $x \ll d$ we expect the field to be that of a point charge $+2q$ at the origin. To test this prediction, we note that for $x \ll d$

$$\frac{x}{(x^2 + d^2)^{3/2}} \approx \frac{x}{d^3} \approx 0 \Rightarrow \vec{E}_{net} = \frac{1}{4\pi\varepsilon_0} \frac{2q}{x^2}\hat{i}$$

This is, indeed, the field on the x-axis of point charge $2q$ at the origin. For $x \gg d$, the three charges appear as a single charge of value $q_{net} = q_1 + q_2 + q_3 = 0$. So we expect $\vec{E} \approx 0$ when $x \gg d$. In this limit,

$$\frac{1}{x^2} - \frac{x}{(x^2 + d^2)^{3/2}} \approx \frac{1}{x^2} - \frac{x}{x^3} = \frac{1}{x^2} - \frac{1}{x^2} = 0$$

so the field does rapidly become zero, as expected.

26.37. Model: The electric field is that of two infinite lines of charge extending out of the page.
Visualize: The line charges lie on the x-axis.
Solve: (a) From Equation 26.15, the electric field strength due to an infinite line of charge at a distance r from the line charge is

$$E = \frac{1}{4\pi\varepsilon_0}\frac{2|\lambda|}{r}$$

For the left and right line charges, the electric fields are

$$E_{\text{left}} = E_{\text{right}} = \frac{1}{4\pi\varepsilon_0}\frac{2\lambda}{\sqrt{y^2+(d/2)^2}} = \frac{1}{4\pi\varepsilon_0}\frac{4\lambda}{\sqrt{4y^2+d^2}}$$

$$\cos\phi = \frac{d/2}{\sqrt{y^2+d^2/4}} = \frac{d}{\sqrt{4y^2+d^2}} \qquad \sin\phi = \frac{2y}{\sqrt{4y^2+d^2}}$$

$$\Rightarrow \vec{E}_{\text{left}} = \frac{1}{4\pi\varepsilon_0}\frac{4\lambda}{\sqrt{4y^2+d^2}}\left(\frac{d}{\sqrt{4y^2+d^2}}\hat{i} + \frac{2y}{\sqrt{4y^2+d^2}}\hat{j}\right)$$

$$\vec{E}_{\text{right}} = \frac{1}{4\pi\varepsilon_0}\frac{4\lambda}{\sqrt{4y^2+d^2}}\left(\frac{d}{\sqrt{4y^2+d^2}}\hat{i} - \frac{2y}{\sqrt{4y^2+d^2}}\hat{j}\right)$$

$$\Rightarrow \vec{E}_{\text{net}} = \vec{E}_{\text{left}} + \vec{E}_{\text{right}} = \frac{1}{4\pi\varepsilon_0}\left(\frac{4\lambda}{\sqrt{4y^2+d^2}}\right)(2)\left(\frac{d}{\sqrt{4y^2+d^2}}\right)\hat{i} = \frac{1}{4\pi\varepsilon_0}\frac{8\lambda d}{(4y^2+d^2)}\hat{i}$$

Thus the field strength is

$$E = \frac{1}{4\pi\varepsilon_0}\frac{8\lambda d}{4y^2+d^2}$$

26.41. Model: The electric field is that of a line charge of length L.
Visualize: Let the bottom end of the rod be the origin of the coordinate system. Divide the rod into many small segments of charge Δq and length $\Delta y'$. Segment i creates a small electric field at the point P that makes an angle θ with the horizontal. The field has both x and y components, but $E_z = 0$ N/C. The distance to segment i from point P is $(x^2+y'^2)^{1/2}$.
Solve: The electric field created by segment i at point P is

$$\vec{E}_i = \frac{\Delta q}{4\pi\varepsilon_0(x^2+y'^2)}(\cos\theta\hat{i} - \sin\theta\,\hat{j}) = \frac{\Delta q}{4\pi\varepsilon_0(x^2+y'^2)}\left(\frac{x}{\sqrt{x^2+y'^2}}\hat{i} - \frac{y'}{\sqrt{x^2+y'^2}}\hat{j}\right)$$

The net field is the sum of all the $\vec{E}_i$, which gives $\vec{E} = \sum_i \vec{E}_i$. Δq is not a coordinate, so before converting the sum to an integral we must relate charge Δq to length $\Delta y'$. This is done through the linear charge density $\lambda = Q/L$, from which we have the relationship

$$\Delta q = \lambda\Delta y' = \frac{Q}{L}\Delta y'$$

With this charge, the sum becomes

$$\vec{E} = \frac{Q/L}{4\pi\varepsilon_0}\sum_i\left[\frac{x\Delta y'}{(x^2+y'^2)^{3/2}}\hat{i} - \frac{y'\Delta y'}{(x^2+y'^2)^{3/2}}\hat{j}\right]$$

Now we let $\Delta y' \rightarrow dy'$ and replace the sum by an integral from $y'=0$ m to $y'=L$. Thus,

$$\vec{E} = \frac{(Q/L)}{4\pi\varepsilon_0}\left(\int_0^L \frac{x\,dy'}{(x^2+y'^2)^{3/2}}\hat{i} - \int_0^L \frac{y'dy'}{(x^2+y'^2)^{3/2}}\hat{j}\right) = \frac{(Q/L)}{4\pi\varepsilon_0}\left(x\left[\frac{y'}{x^2\sqrt{x^2+y'^2}}\right]_0^L \hat{i} - \left[\frac{-1}{\sqrt{x^2+y'^2}}\right]_0^L \hat{j}\right)$$

$$= \frac{Q}{4\pi\varepsilon_0}\frac{1}{x\sqrt{x^2+L^2}}\hat{i} - \frac{1}{4\pi\varepsilon_0}\left(\frac{Q}{Lx}\right)\left(1 - \frac{x}{\sqrt{x^2+L^2}}\right)\hat{j}$$

26.43. Model: Assume that the ring of charge is thin and that the charge lies along circle of radius R.
Solve: (a) From Example 26.4, the on-axis field of a ring of charge Q and radius R is

$$E_z = \frac{1}{4\pi\varepsilon_0}\frac{zQ}{(z^2+R^2)^{3/2}}$$

For the field to be maximum at a particular value of z, $dE/dz = 0$. Taking the derivative,

$$\frac{dE}{dz} = \frac{Q}{4\pi\varepsilon_0}\left[\frac{1}{(z^2+R^2)^{3/2}} - \frac{z(3/2)(2z)}{(z^2+R^2)^{5/2}}\right] = 0 \Rightarrow \frac{1}{(z^2+R^2)^{3/2}} = \frac{3z^2}{(z^2+R^2)^{5/2}}$$

$$\Rightarrow 1 = \frac{3z^2}{z^2+R^2} \Rightarrow z = \pm\frac{R}{\sqrt{2}}$$

(b) The field strength at the point $z = R/\sqrt{2}$ is

$$(E_z)_{max} = \frac{Q}{4\pi\varepsilon_0}\frac{(R/\sqrt{2})}{\left[(R/\sqrt{2})^2+R^2\right]^{3/2}} = \frac{2}{3\sqrt{3}}\frac{Q}{4\pi\varepsilon_0 R^2}$$

26.47. Model: An insulating sphere of charge Q and radius R has an electric field *outside* the sphere that is exactly the same as that of a point charge Q located at the center of the sphere.
Visualize:

$Q_1 = 10$ nC $Q_2 = -15$ nC

Solve: The electric field of a charged sphere at a distance $r > R$ is given by Equation 26.28:

$$\vec{E}_{sphere} = \frac{Q}{4\pi\varepsilon_0 r^2}\hat{r}$$

In the present case, $\vec{E} = \vec{E}_1 + \vec{E}_2$, where $\vec{E}_1$ and $\vec{E}_2$ are the fields of the individual spheres. The distance from the center of each sphere to the midpoint between is $r_1 = r_2 = 4$ cm. Thus,

$$E_1 = \frac{(9.0\times10^9 \text{ N m}^2/\text{C}^2)(10\times10^{-9}\text{ C})}{(0.040\text{ m})^2} = 5.625\times10^4 \text{ N/C}$$

$$E_2 = \frac{(9.0\times10^9 \text{ N m}^2/\text{C}^2)(15\times10^{-9}\text{ C})}{(0.040\text{ m})^2} = 8.438\times10^4 \text{ N/C}$$

The fields point in the same direction, so

$$\vec{E} = (5.625\times10^4 \text{ N/C} + 8.438\times10^4 \text{ N/C, right}) = (1.41\times10^5 \text{ N/C, right}).$$

The electric field will point left when Q_1 and Q_2 are interchanged. The electric field strength in both cases is 1.41×10^5 N/C.

26.49. Model: The electric field is uniform inside the capacitor, so constant-acceleration kinematic equations apply to the motion of the proton.
Visualize:

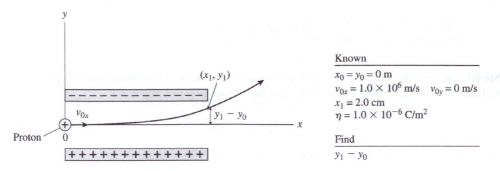

Known
$x_0 = y_0 = 0$ m
$v_{0x} = 1.0 \times 10^6$ m/s $v_{0y} = 0$ m/s
$x_1 = 2.0$ cm
$\eta = 1.0 \times 10^{-6}$ C/m^2

Find
$y_1 - y_0$

Solve: From Equation 26.29, the electric field between the parallel plates $\vec{E} = (\eta/\varepsilon_0)\hat{j}$. The force on the proton is

$$\vec{F} = m\vec{a} = q\vec{E} \Rightarrow \vec{a} = \frac{q\vec{E}}{m} = \frac{q\eta}{m\varepsilon_0}\hat{j} \Rightarrow a_y = \frac{q\eta}{m\varepsilon_0}$$

Using the kinematic equation $y_1 = y_0 + v_{0y}(t_1 - t_0) + \frac{1}{2}a_y(t_1 - t_0)^2$,

$$\Delta y = y_1 - y_0 = (0 \text{ m/s})(t_1 - t_0) + \frac{1}{2}a_y(t_1 - t_0) = \frac{1}{2}\left(\frac{q\eta}{m\varepsilon_0}\right)(t_1 - t_0)^2$$

To determine $t_1 - t_0$, we consider the horizontal motion of the proton. The proton travels a distance of 2.0 cm at a constant speed of 1.0×10^6 m/s. The velocity is constant because the only force acting on the proton is due to the field between the plate along the y-direction. Using the same kinematic equation,

$$\Delta x = 2.0 \times 10^{-2} \text{ m} = v_{0x}(t_1 - t_0) + 0 \text{ m} \Rightarrow (t_1 - t_0) = \frac{2.0 \times 10^{-2} \text{ m}}{1.0 \times 10^6 \text{ m/s}} = 2.0 \times 10^{-8} \text{ s}$$

$$\Rightarrow \Delta y = \frac{1}{2}\frac{(1.60 \times 10^{-19} \text{ C})(1.0 \times 10^{-6} \text{ C/m}^2)(2.0 \times 10^{-8} \text{ s})^2}{(1.67 \times 10^{-27} \text{ kg})(8.85 \times 10^{-12} \text{ C}^2/\text{Nm}^2)} = 2.2 \text{ mm}$$

26.53. Model: Assume the soot particle is a point charge in $F = qE$ and a small sphere where the drag force is concerned. Also assume the particle becomes negatively charged.
Visualize: Draw a free body diagram with the electric force directed down because the E field is down, but the actual direction of that force depends on the sign of q.

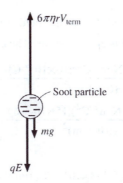

Solve: (a) Write Newton's first law using the free body diagram.

$$\Sigma F = 6\pi\eta r v_{\text{term}} - mg - qE = 0$$

Note that $m = V\rho = \frac{4}{3}\pi r^3 \rho$. Solve for v_{term}.

$$v_{term} = \frac{\left(\frac{4}{3}\pi r^3 \rho\right)g + qE}{6\pi\eta r}$$

(b) With the qE term equal to zero the π and an r will cancel.

$$v_{term} = \frac{\frac{4}{3}r^2 \rho g}{6\eta} = \frac{\frac{4}{3}(0.50 \ \mu m)^2 (2200 \ kg/m^3)(9.8 \ m/s^2)}{6(1.8\times10^{-5} \ kg/m\cdot s)} = 0.067 \ mm/s$$

(c) The charge on the soot particle is negative, so the electric force is up.

$$v_{term} = \frac{\frac{4}{3}\pi(0.50 \ \mu m)^3 (2200 \ kg/m^3)(9.8 \ m/s^2) + (125)(-1.60\times10^{-19} \ C)(150 \ N/C)}{6\pi(1.8\times10^{-5} \ kg/m\cdot s)(0.50 \ \mu m)} = 0.049 \ mm/s$$

Assess: The terminal speed is smaller when the electric force is present, as we would expect.

26.55. Model: The electron orbiting the proton experiences a force given by Coulomb's law.
Visualize:

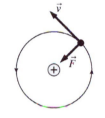

Solve: The force that causes the circular motion is

$$F = \frac{1}{4\pi\varepsilon_0}\frac{q_p|q_e|}{r^2} = \frac{m_e v^2}{r} = \frac{m_e(4\pi^2 r^2 f^2)}{r}$$

where we used $v = 2\pi r/T = 2\pi rf$. The frequency is

$$f = \sqrt{\left(\frac{1}{4\pi\varepsilon_0}\right)\frac{q_p|q_e|}{4\pi^2 m_e r^3}} = \sqrt{\frac{(9.0\times10^9 \ N\,m^2/C^2)(1.60\times10^{-19} \ C)^2}{4\pi^2(9.11\times10^{-31} \ kg)(5.3\times10^{-11} \ m)^3}} = 6.56\times10^{15} \ Hz$$

26.57. Model: The electric field at the dipole's location is that of the ion with charge q.
Visualize:

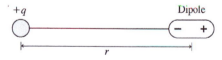

Solve: (a) We have $p = \alpha E$. The units of α are the units of p/E and are

$$\frac{C\,m}{N/C} = \frac{C^2\,m}{N} = \frac{C^2 s^2}{kg}$$

(b) The electric field due to the ion at the location of the dipole is

$$E_{at\ dipole} = \left(\frac{1}{4\pi\varepsilon_0}\frac{q}{r^2}, \text{ away from } q\right) = \frac{1}{4\pi\varepsilon_0}\frac{q}{r^2}\hat{i}$$

Because $\vec{p} = \alpha \vec{E}$, the induced dipole moment is

$$\vec{p} = \alpha \left(\frac{1}{4\pi\varepsilon_0} \frac{q}{r^2} \right) \hat{i}$$

From Equation 26.11, the electric field produced by the dipole at the location of the ion is

$$\vec{E}_{\text{dipole}} = \frac{1}{4\pi\varepsilon_0} \frac{2\vec{p}}{r^3} = \frac{1}{4\pi\varepsilon_0} \left(\frac{2}{r^3} \right) \alpha \left(\frac{1}{4\pi\varepsilon_0} \frac{q}{r^2} \right) \hat{i} = \left(\frac{1}{4\pi\varepsilon_0} \right)^2 \left(\frac{2q\alpha}{r^5} \right) \hat{i}$$

The force the dipole exerts on the ion is

$$\vec{F}_{\text{dipole on ion}} = q\vec{E}_{\text{dipole}} = \left(\frac{1}{4\pi\varepsilon_0} \right)^2 \left(\frac{2q^2\alpha}{r^5} \right) \hat{i}$$

According to Newton's third law, $\vec{F}_{\text{dipole on ion}} = -\vec{F}_{\text{ion on dipole}}$. Therefore,

$$\vec{F}_{\text{ion on dipole}} = \left(\left(\frac{1}{4\pi\varepsilon_0} \right)^2 \frac{2q^2\alpha}{r^5}, \text{ toward ion} \right)$$

GAUSS'S LAW

Exercises and Problems

Section 27.1 Symmetry

27.3. Visualize:

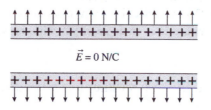

$$\vec{E} = 0 \text{ N/C}$$

Figure 27.6 shows the electric field for an infinite plane of charge. For two parallel planes, this is the only shape of the electric field vectors that matches the symmetry of the charge distribution. Because the charge density is equal on each plane, there can be no electric field between the planes.

Section 27.2 The Concept of Flux

27.5. Model: The electric flux "flows" *out* of a closed surface around a region of space containing a net positive charge and *into* a closed surface surrounding a net negative charge.

Visualize: Please refer to Figure EX27.5. Let A be the area of each of the six faces of the cube and let the direction of $\vec{A}$ be outward from the surfaces.

Solve: The electric flux is defined as $\Phi_e = \vec{E} \cdot \vec{A} = EA\cos\theta$, where θ is the angle between the electric field and a line *perpendicular* to the plane of the surface (i.e., the "normal" to the surface). The electric flux out of the closed cube surface is

$$\Phi_{out} = (10 \text{ N/C} + 10 \text{ N/C} + 20 \text{ N/C})A\cos(0°) = (40A) \text{ N m}^2/\text{C}$$

Similarly, the electric flux into the closed cube surface is

$$\Phi_{in} = (15 \text{ N/C} + 20 \text{ N/C} + 5 \text{ N/C})A\cos(180°) = -(40A) \text{ N m}^2/\text{C}$$

Thus, $\Phi_{out} + \Phi_{in} = 0 \text{ N m}^2/\text{C}$. Since there is no net electric flux passing through the surface of the closed box, it contains no charge.

27.7. Model: The electric flux "flows" *out* of a closed surface around a region of space containing a net positive charge and *into* a closed surface surrounding a net negative charge.

Visualize: Please refer to Figure EX27.7. Let A be the area of each of the six faces of the cube and let the direction of $\vec{A}$ be outward from the surfaces.

Solve: The electric flux is defined as $\Phi_e = \vec{E} \cdot \vec{A} = EA\cos\theta$, where θ is the angle between the electric field vector and an outward unit vector that is perpendicular to the plane of the surface. The electric flux out of the closed cube surface is

$$\Phi_{out} = (20 \text{ N/C} + 10 \text{ N/C} + 10 \text{ N/C})A\cos(0°) = (40A) \text{ N m}^2/\text{C}$$

Similarly, the electric flux into the closed cube surface is

$$\Phi_{in} = (20 \text{ N/C} + 15 \text{ N/C})A\cos(180°) = -(35A) \text{ N m}^2/\text{C}$$

Because the cube contains negative charge, $\Phi_{out} + \Phi_{in}$ must be negative. This means $\Phi_{out} + \Phi_{in} + \Phi_{unknown} < 0 \text{ N m}^2/\text{C}$. Therefore,

$$(40A) \text{ N m}^2/\text{C} + (-35A) \text{ N m}^2/\text{C} + \Phi_{unknown} < 0 \text{ N m}^2/\text{C}$$

$$\Phi_{unknown} < (-5A) \text{ N m}^2/\text{C}$$

That is, the unknown electric field vector can point either *into* or *out of* the front face of the cube, provided its field strength is greater than −5 N/C.

Section 27.3 Calculating Electric Flux

27.9. Model: The electric field is uniform over the entire surface.
Visualize: Please refer to Figure EX27.9. The electric field vectors make an angle of 30° with the planar surface. Because the normal $\hat{n}$ to the planar surface is at an angle of 90° with the surface, the angle between $\hat{n}$ and $\vec{E}$ is $\theta = 60°$.
Solve: The electric flux is

$$\Phi_e = \vec{E} \cdot \vec{A} = EA\cos\theta = (200 \text{ N/C})(1.0 \times 10^{-2} \text{ m}^2)\cos(60°) = 1.0 \text{ N m}^2/\text{C}$$

27.15. Model: The electric field is uniform, and we take the area vectors to point outward from the box.
Visualize: In the figure below, the box is positioned with its edges aligned with the *xyz* axes, and the electric field is evaluated at the input face and the exit face.

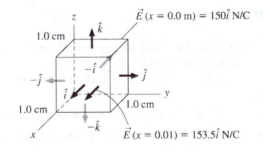

Solve: The area vectors of the six box faces are $\vec{A}_1 = (1.0 \times 10^{-2} \text{ m})^2 \hat{i} = (1.0 \times 10^{-4} \text{ m}^2)\hat{i}$, $\vec{A}_2 = -(1.0 \times 10^{-4} \text{ m}^2)\hat{i}$, $\vec{A}_3 = (1.0 \times 10^{-4} \text{ m}^2)\hat{j}$, $\vec{A}_4 = -(1.0 \times 10^{-4} \text{ m}^2)\hat{j}$, $\vec{A}_5 = (1.0 \times 10^{-4} \text{ m}^2)\hat{k}$, and $\vec{A}_6 = -(1.0 \times 10^{-4} \text{ m}^2)\hat{k}$. The net flux through the box is

$$\Phi = \vec{E} \cdot \vec{A} = \sum_{i=1}^{6} \vec{E} \cdot \vec{A}_i = \vec{E}(x = 0.01 \text{ m}) \cdot \vec{A}_1 + \vec{E}(x = 0.0 \text{ m}) \cdot \vec{A}_2$$

$$= (153.5 \text{ N/C})(1.0 \times 10^{-4} \text{ m}^2) - (150 \text{ N/C})(1.0 \times 10^{-4} \text{ m}^2)$$

$$= 3.5 \times 10^{-4} \text{ N m}^2/\text{C}$$

Section 27.4 Gauss's Law

Section 27.5 Using Gauss's Law

27.17. Visualize:

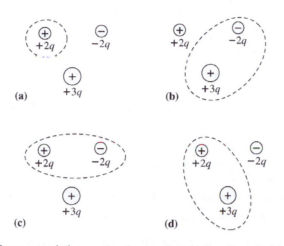

(a)

(b)

(c)

(d)

For *any* closed surface that encloses a total charge Q_{in}, the net electric flux through the closed surface is $\Phi_e = Q_{in}/\epsilon_0$.

27.19. Visualize: Please refer to Figure EX27.19.
Solve: For *any* closed surface that encloses a total charge Q_{in}, the net electric flux through the surface is $\Phi_e = Q_{in}/\epsilon_0$. We can write three equations from the three closed surfaces in the figure:

$$\Phi_A = -\frac{q}{\epsilon_0} = \frac{q_1 + q_3}{\epsilon_0} \implies q_1 + q_3 = -q \qquad \Phi_B = \frac{3q}{\epsilon_0} = \frac{q_1 + q_2}{\epsilon_0} \implies q_1 + q_2 = 3q$$

$$\Phi_C = \frac{-2q}{\epsilon_0} = \frac{q_2 + q_3}{\epsilon_0} \implies q_2 + q_3 = -2q$$

Subtracting third equation from the first gives

$$q_1 - q_2 = +q$$

Adding second equation to this equation,

$$2q_1 = +4q \quad q_1 = 2q$$

That is, $q_1 = +2q$, $q_2 = +q$, and $q_3 = -3q$.

27.21. Visualize: Please refer to Figure EX27.21. For *any* closed surface that encloses a total charge Q_{in}, the net electric flux through the closed surface is $\Phi_e = Q_{in}/\epsilon_0$. The cylinder encloses the +1 nC charge only as both the +100 nC and the −100 nC charges are outside the cylinder. Thus,

$$\Phi_e = \frac{1 \times 10^{-9} \text{ C}}{8.85 \times 10^{-12} \text{ C}^2/\text{Nm}^2} = 0.11 \text{ kN m}^2/\text{C}$$

This is outward flux.

Section 27.6 Conductors in Electrostatic Equilibrium

27.27. Model: The copper plate is a conductor. The excess charge resides on the surface of the plate. Ignore the charge that resides on the edge of the plate because the plate's thickness is much, much less than the radius.
Solve: (a) One-half of the charge is located on the top surface and one-half on the bottom surface of the copper plate, so the surface charge density is

$$\eta = \frac{q}{A} = \frac{(3.5/2 \text{ nC})}{\pi(0.10/2 \text{ m})^2} = 2.23 \times 10^{-7} \text{ C/m}^2$$

Thus, the electric field at the surface of the plate is

$$E = \frac{\eta}{\epsilon_0} = \frac{2.23 \times 10^{-7} \text{ C/m}^2}{8.85 \times 10^{-12} \text{ C}^2/\text{Nm}^2} = 2.52 \times 10^4 \text{ N/C}$$

Because the charge on the plate is positive, the direction of the electric field is away from the plate. Thus the electric field is $\vec{E} = 25 \text{ kN/C}$ upward from the plate.

(b) The center of mass of the plate is in the interior of the plate, so $E = 0.0 \text{ N/C}$ because the electric field within a conductor is zero.

(c) The electric field $E = 25 \times 10^3 \text{ N/C}$, away from the plate, which is downward. Thus $\vec{E} = 25 \text{ kN/C}$ downward from the plate.

27.29. Model: The electric field over the four faces of the cube is uniform.
Visualize: Please refer to Figure P27.29.
Solve: (a) The electric flux through a surface area $\vec{A}$ is $\Phi = \vec{E} \cdot \vec{A} = EA\cos\theta$, where θ is the angle between the electric field and the vector $\vec{A}$ that points *outward* from the surface. Thus

$$\Phi_1 = EA\cos\theta_1 = (500 \text{ N/C})(9.0 \times 10^{-4} \text{ m}^2)\cos(150°) = -0.39 \text{ N m}^2/\text{C}$$

$$\Phi_2 = EA\cos\theta_2 = (500 \text{ N/C})(9.0 \times 10^{-4} \text{ m}^2)\cos(60°) = 0.23 \text{ N m}^2/\text{C}$$

$$\Phi_3 = EA\cos\theta_3 = (500 \text{ N/C})(9.0 \times 10^{-4} \text{ m}^2)\cos(30°) = 0.39 \text{ N m}^2/\text{C}$$

$$\Phi_4 = EA\cos\theta_4 = (500 \text{ N/C})(9.0 \times 10^{-4} \text{ m}^2)\cos(120°) = -0.23 \text{ N m}^2/\text{C}$$

(b) The net flux through these four sides is $\Phi_T = \Phi_1 + \Phi_2 + \Phi_3 + \Phi_4 = 0.0 \text{ N m}^2/\text{C}$.

Assess: The net flux through the four faces of the cube is zero because there is no enclosed charge.

27.33. Solve: For *any* closed surface that encloses a total charge Q_{in}, the net electric flux through the closed surface is $\Phi_e = Q_{in}/\epsilon_0$. The flux through the top surface of the cube is one-sixth of the total:

$$\Phi_{e \text{ surface}} = \frac{Q_{in}}{6\epsilon_0} = \frac{10 \times 10^{-9} \text{ C}}{6(8.85 \times 10^{-12} \text{ C}^2/\text{Nm}^2)} = 0.19 \text{ kN m}^2/\text{C}$$

27.35. Solve: (a) The electric field is

$$\vec{E} = \left(\frac{200}{0.10}\right)\hat{r} \text{ N/C} = 2.0\hat{r} \text{ kN/C}$$

So the electric field strength is 2.0 kN/C.

(b) The area of the spherical surface is $A_{\text{sphere}} = 4\pi(0.10 \text{ m})^2 = 0.1257 \text{ m}^2$. Hence, the flux is

$$\Phi_e = \oint \vec{E} \cdot d\vec{A} = EA_{\text{sphere}} = (2000 \text{ N/C})(0.1257 \text{ m}^2) = 0.25 \text{ kN m}^2/\text{C}$$

(c) Because $\Phi_e = Q_{in}/\epsilon_0$,

$$Q_{in} = \epsilon_0 \Phi_e = (8.85 \times 10^{-12} \text{ C}^2/\text{Nm}^2)(250 \text{ N m}^2/\text{C}) = 2.2 \times 10^{-9} \text{ C} = 2.2 \text{ nC}$$

27.37. Model: The excess charge on a conductor resides on the outer surface.
Visualize:

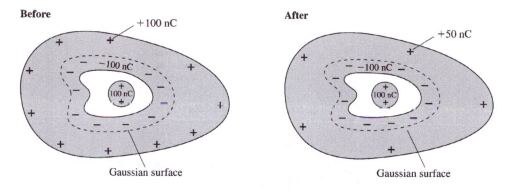

Before   +100 nC

After   +50 nC

Gaussian surface   Gaussian surface

Solve: (a) Consider a Gaussian surface surrounding the cavity just *inside* the conductor. The electric field inside a conductor in electrostatic equilibrium is zero, so $\vec{E}$ is zero at all points on the Gaussian surface. Thus $\Phi_e = 0$. Gauss's law tells us that $\Phi_e = Q_{in}/\epsilon_0$, so the net charge enclosed by this Gaussian surface is $Q_{in} = Q_{point} + Q_{wall} = 0$. We know that $Q_{point} = +100$ nC, so $Q_{wall} = -100$ nC. The positive charge in the cavity attracts an equal negative charge to the inside surface.
(b) The conductor started out neutral. If there is -100 nC on the wall of the cavity, then the exterior surface of the conductor was initially $+100$ nC. Transferring -50 nC to the conductor reduces the exterior surface charge by 50 nC, leaving it at $+50$ nC.
Assess: The electric field inside the conductor stays zero.

27.41. Model: The charge distribution at the surface of the earth is assumed to be uniform and to have spherical symmetry.
Visualize:

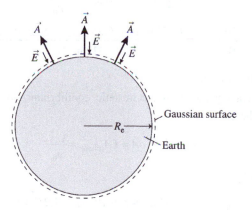

Gaussian surface

R_e

Earth

Due to the symmetry of the charge distribution, $\vec{E}$ is perpendicular to the Gaussian surface and the field strength has the same value at all points on the surface.
Solve: Gauss's law is $\Phi_e = \oint \vec{E} \cdot d\vec{A} = Q_{in}/\epsilon_0$. The electric field points inward (negative flux), hence

$$Q_{in} = -\epsilon_0 E A_{sphere} = -(8.85 \times 10^{-12} \text{ C}^2/\text{Nm}^2)(100 \text{ N/C})4\pi(6.37 \times 10^6 \text{ m})^2 = -4.51 \times 10^5 \text{ C}$$

27.43. Model: The hollow metal sphere is charged such that the charge distribution is spherically symmetric.
Visualize:

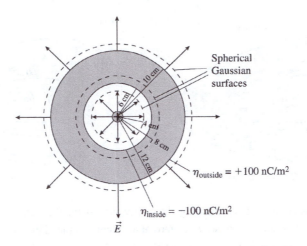

The figure shows spherical Gaussian surfaces at $r = 4$ cm, at $r = 8$ cm, and at $r = 12$ cm. These surfaces match the symmetry of the spherical charge distribution, so $\vec{E}$ is perpendicular to the Gaussian surface and the field strength has the same value at all points on a given Gaussian surface.
Solve: The charge on the inside surface is

$$Q_{\text{inside}} = (-100 \text{ nC/m}^2) \; 4\pi(0.06 \text{ m})^2 = -4.524 \text{ nC}$$

This charge is caused by polarization. That is, the inside surface can be charged only if there is a charge of +4.524 nC at the center that polarizes the metal sphere. Applying Gauss's law to the 4.0-cm-radius Gaussian surface, which encloses the +4.524 nC charge gives

$$\Phi_e = \oint \vec{E} \cdot d\vec{A} = EA_{\text{sphere}} = \frac{Q_{\text{in}}}{\varepsilon_0}$$

$$E = \frac{Q_{\text{in}}}{A\varepsilon_0} = \frac{4.524 \times 10^{-9} \text{ C}}{4\pi\varepsilon_0 (0.040 \text{ m})^2} = \frac{(4.524 \times 10^{-9} \text{ C})(9.0 \times 10^9 \text{ N m}^2/\text{C}^2)}{(0.040 \text{ m})^2} = 2.54 \times 10^4 \text{ N/C}$$

Thus, at $r = 4.0$ cm, $\vec{E} = (2.5 \times 10^4 \text{ N/C, outward})$.
There is no electric field inside a conductor in electrostatic equilibrium. So at $r = 8$ cm, $E = 0$ N/C. Applying Gauss's law to a 12-cm-radius sphere,

$$\Phi_e = \oint \vec{E} \cdot d\vec{A} = EA_{\text{sphere}} = \frac{Q_{\text{in}}}{\varepsilon_0}$$

The charge on the outside surface is

$$Q_{\text{outside}} = (+100 \text{ nC/m}^2) 4\pi(0.10 \text{ m})^2 = 1.257 \times 10^{-8} \text{ C}$$

$$Q_{\text{in}} = Q_{\text{outside}} + Q_{\text{inside}} + Q_{\text{center}} = 1.257 \times 10^{-8} \text{ C} - 0.452 \times 10^{-8} \text{ C} + 0.452 \times 10^{-8} \text{ C} = 1.257 \times 10^{-8} \text{ C}$$

$$E = \frac{Q_{\text{in}}}{\varepsilon_0 A} = \frac{1.257 \times 10^{-8} \text{ C}}{(8.85 \times 10^{-12} \text{ C}^2/\text{Nm}^2) 4\pi(0.12 \text{ m})^2} = 7.86 \times 10^3 \text{ N/C}$$

At $r = 12$ cm, $\vec{E} = (7.9 \times 10^3 \text{ N/C, outward})$.

27.45. Model: The hollow plastic ball has a charge uniformly distributed on its outer surface. This distribution leads to a spherically symmetric electric field. Assume $Q > 0$.

Visualize:

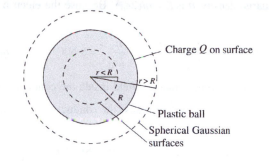

The figure shows Gaussian surfaces at $r < R$ and $r > R$.

Solve: **(a)** Gauss's law for the Gaussian surface for $r < R$ where $Q_{in} = 0$ is

$$\Phi_e = \oint \vec{E} \cdot d\vec{A} = \frac{Q_{in}}{\varepsilon_0} = 0 \text{ N m}^2/\text{C} \quad \Rightarrow \quad E = 0 \text{ N/C}$$

(b) Gauss's law for the Gaussian surface for $r > R$ is

$$\Phi_e = \oint \vec{E} \cdot d\vec{A} = EA_{sphere} = \frac{Q_{in}}{\varepsilon_0} = \frac{Q}{\varepsilon_0} \quad \Rightarrow \quad EA_{sphere} = \frac{Q}{\varepsilon_0} \quad \Rightarrow \quad E = \frac{Q}{\varepsilon_0 A_{sphere}} = \frac{1}{4\pi\varepsilon_0}\frac{Q}{r^2}$$

Because $Q > 0$, the electric field points radially outward, so $\vec{E} = \frac{1}{4\pi\varepsilon_0}\frac{Q}{r^2}\hat{r}$

Assess: A uniform spherical shell of charge has the same electric field at $r > R$ as a point charge placed at the center of the sphere. Additionally, the shell of charge exerts no electric force on a charged particle inside the shell.

27.49. Model: The infinitely wide plane of charge with surface charge density η polarizes the infinitely wide conductor.
Visualize:

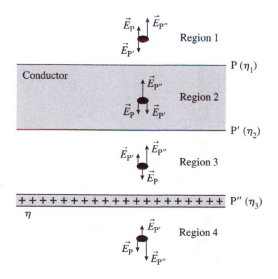

Because $\vec{E} = \vec{0}$ in the metal there will be an induced charge polarization. The face of the conductor adjacent to the plane of charge is negatively charged. This makes the other face of the conductor positively charged. We thus have three infinite planes of charge. These are P (top conducting face), P' (bottom conducting face), and P'' (plane of charge).

Solve: Let η_1, η_2, and η_3 be the surface charge densities of the three surfaces with $\eta_2 < 0$. The electric field due to a plane of charge with surface charge density η is $E = \eta/(2\epsilon_0)$. Because the electric field inside a conductor is zero (region 2),

$$\vec{E}_P + \vec{E}_{P'} + \vec{E}_{P''} = \vec{0} \text{ N/C} \quad \Rightarrow \quad -\frac{\eta_1}{2\epsilon_0}\hat{j} + \frac{\eta_2}{2\epsilon_0}\hat{j} + \frac{\eta_3}{2\epsilon_0}\hat{j} = \vec{0} \text{ N/C} \quad \Rightarrow \quad -\eta_1 + \eta_2 + \eta = 0 \text{ C/m}^2$$

We have made the substitution $\eta_3 = \eta$. Also note that the field inside the conductor is downward from planes P and P' and upward from P''. Because $\eta_1 + \eta_2 = 0 \text{ C/m}^2$, because the conductor is neutral, $\eta_2 = -\eta_1$. The above equation becomes

$$-\eta_1 - \eta_1 + \eta = 0 \text{ C/m}^2 \quad \Rightarrow \quad \eta_1 = \tfrac{1}{2}\eta \Rightarrow \eta_2 = -\tfrac{1}{2}\eta$$

We are now in a position to find electric field in regions 1–4.
For region 1,

$$\vec{E}_P = \frac{\eta}{4\epsilon_0}\hat{j} \qquad \vec{E}_{P'} = -\frac{\eta}{4\epsilon_0}\hat{j} \qquad \vec{E}_{P''} = \frac{\eta}{2\epsilon_0}\hat{j}$$

The electric field is $\vec{E}_{net} = \vec{E}_P + \vec{E}_{P'} + \vec{E}_{P''} = [\eta/(2\epsilon_0)]\hat{j}$.
In region 2, $\vec{E}_{net} = \vec{0}$ N/C. In region 3,

$$\vec{E}_P = -\frac{\eta}{4\epsilon_0}\hat{j} \qquad \vec{E}_{P'} = \frac{\eta}{4\epsilon_0}\hat{j} \qquad \vec{E}_{P''} = \frac{\eta}{2\epsilon_0}\hat{j}$$

The electric field is $\vec{E}_{net} = [\eta/(2\epsilon_0)]\hat{j}$.
In region 4,

$$\vec{E}_P = -\frac{\eta}{4\epsilon_0}\hat{j} \qquad \vec{E}_{P'} = \frac{\eta}{4\epsilon_0}\hat{j} \qquad \vec{E}_{P''} = -\frac{\eta}{2\epsilon_0}\hat{j}$$

The electric field is $\vec{E}_{net} = -[\eta/(2\epsilon_0)]\hat{j}$.

27.51. Model: A long, charged wire can be modeled as an infinitely long line of charge.
Visualize:

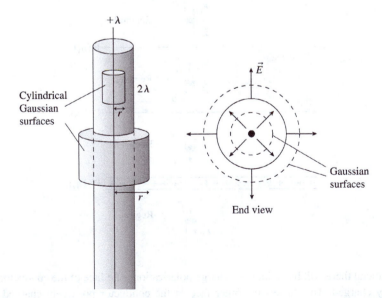

The figure shows an infinitely long line of charge that is surrounded by a hollow metal cylinder of radius R. The symmetry of the situation indicates that the only possible shape of the electric field is to point straight in or out from the wire. The shape of the field suggests that we choose our Gaussian surface to be a cylinder of radius r and length L, centered on the wire.

Solve: **(a)** For the region $r < R$, Gauss's law is

$$\Phi_e = \oint \vec{E} \cdot d\vec{A} = \frac{Q_{in}}{\epsilon_0} \quad \Rightarrow \quad \int_{top} \vec{E} \cdot d\vec{A} + \int_{bottom} \vec{E} \cdot d\vec{A} + \int_{side} \vec{E} \cdot d\vec{A} = \frac{\lambda L}{\epsilon_0}$$

$$0 \text{ N m}^2/\text{C} + 0 \text{ N m}^2/\text{C} + \vec{E} \cdot \vec{A}_{side} = \frac{\lambda L}{\epsilon_0} \quad \Rightarrow \quad E(2\pi r)L = \frac{\lambda L}{\epsilon_0}$$

$$E = \frac{\lambda}{2\pi\epsilon_0} \frac{1}{r} \quad \Rightarrow \quad \vec{E} = \left(\frac{\lambda}{2\pi\epsilon_0} \frac{1}{r}, \text{ outward} \right) = \frac{\lambda}{2\pi\epsilon_0} \frac{\hat{r}}{r}$$

(b) Applying Gauss's law to the Gaussian surface at $r > R$,

$$\oint \vec{E} \cdot d\vec{A} = \int_{top} \vec{E} \cdot d\vec{A} + \int_{bottom} \vec{E} \cdot d\vec{A} + \int_{side} \vec{E} \cdot d\vec{A} = \frac{Q_{in}}{\epsilon_0}$$

$$0 \text{ N m}^2/\text{C} + 0 \text{ N m}^2/\text{C} + \vec{E} \cdot \vec{A}_{wall} = \frac{Q_{in}}{\epsilon_0}$$

$$E(2\pi rL) = \frac{Q_{line} + Q_{cylinder}}{\epsilon_0} = \frac{\lambda L + 2\lambda L}{\epsilon_0} \quad \Rightarrow \quad E = \frac{3\lambda}{2\pi\epsilon_0} \frac{1}{r} \quad \Rightarrow \quad \vec{E} = \frac{3\lambda}{2\pi\epsilon_0} \frac{\hat{r}}{r}$$

THE ELECTRIC POTENTIAL

Exercises and Problems

Section 28.1 Electric Potential Energy

28.1. Model: The mechanical energy of the proton is conserved. A parallel-plate capacitor has a uniform electric field.
Visualize:

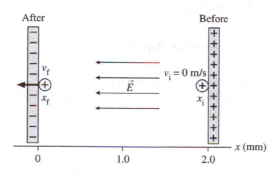

The figure shows the before-and-after pictorial representation. The proton has an initial speed $v_i = 0$ m/s and a final speed v_f after traveling a distance $d = 2.0$ mm.

Solve: The proton loses potential energy and gains kinetic energy as it moves toward the negative plate. The potential energy U is defined as $U = U_0 + qEx$, where x is the distance from the negative plate and U_0 is the potential energy at the negative plate (at $x = 0$ m). Thus, the change in the potential energy of the proton is

$$\Delta U_p = U_f - U_i = (U_0 + 0 \text{ J}) - (U_0 + qEd) = -qEd$$

The change in the kinetic energy of the proton is

$$\Delta K = K_f - K_i = \tfrac{1}{2}mv_f^2 - \tfrac{1}{2}mv_i^2 = \tfrac{1}{2}mv_f^2$$

The law of conservation of energy is $\Delta K + \Delta U_p = 0$ J. This means

$$\tfrac{1}{2}mv_f^2 + (-qEd) = 0 \text{ J}$$

$$\Rightarrow v_f = \sqrt{\frac{2qEd}{m}} = \sqrt{\frac{2(+1.60\times10^{-19} \text{ C})(50{,}000 \text{ N/C})(2.0\times10^{-3} \text{ m})}{(1.67\times10^{-27} \text{ kg})}} = 1.4\times10^5 \text{ m/s}$$

Assess: As described in Section 28.1, the potential energy for a charge q in an electric field E is $U = U_0 + qEx$, where x is the distance measured from the negative plate. Having $U = U_0$ at the negative plate (with $x = 0$ m) is completely arbitrary. We could have taken it to be zero. Note that only ΔU, and not U, has physical consequences.

Section 28.2 The Potential Energy of Point Charges

28.5. Model: The charges are point charges.
Solve: The electric potential energy of the proton is

$$U_{\text{proton}} = U_{13} + U_{23}$$

$$= (9.0 \times 10^9 \text{ N m}^2/\text{C}^2) \left[\frac{(1.60 \times 10^{-19} \text{ C})(-1.60 \times 10^{-19} \text{ C})}{\sqrt{(2.0 \times 10^{-9} \text{ m})^2 + (0.50 \times 10^{-9} \text{ m})^2}} + \frac{(1.60 \times 10^{-19} \text{ C})(-1.60 \times 10^{-19} \text{ C})}{\sqrt{(2.0 \times 10^{-9} \text{ m})^2 + (0.50 \times 10^{-9} \text{ m})^2}} \right]$$

$$= -1.12 \times 10^{-19} \text{ J} - 1.12 \times 10^{-19} \text{ J} = -2.24 \times 10^{-19} \text{ J} \approx -2.2 \times 10^{-19} \text{ J}$$

28.7. Model: The charges are point charges.
Solve: For a system of point charges, the potential energy is the sum of the potential energies due to all distinct pairs of charges:

$$U_{\text{elec}} = \sum_{i,j} \frac{Kq_i q_j}{r_{ij}} = U_{12} + U_{13} + U_{23}$$

$$= (9.0 \times 10^9 \text{ N m}^2/\text{C}^2) \left[\frac{(2.0 \times 10^{-9} \text{ C})(3.0 \times 10^{-9} \text{ C})}{0.030 \text{ m}} + \frac{(2.0 \times 10^{-9} \text{ C})(3.0 \times 10^{-9} \text{ C})}{0.040 \text{ m}} + \frac{(3.0 \times 10^{-9} \text{ C})(3.0 \times 10^{-9} \text{ C})}{\sqrt{(0.030 \text{ m})^2 + (0.040 \text{ m})^2}} \right]$$

$$= 1.80 \times 10^{-6} \text{ J} + 1.35 \times 10^{-6} \text{ J} + 1.62 \times 10^{-6} \text{ J} = 4.8 \times 10^{-6} \text{ J}$$

Assess: Note that $U_{12} = U_{21}$, $U_{13} = U_{31}$, and $U_{23} = U_{32}$.

Section 28.3 The Potential Energy of a Dipole

28.9. Model: An external electric field supplies energy to a dipole.
Visualize: On an energy diagram, the oscillation occurs between the points where the potential-energy curve crosses the total energy line.

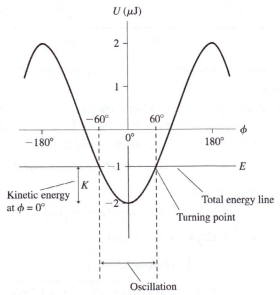

Solve: (a) The potential energy of an electric dipole moment in a uniform electric field is $U = -\vec{p} \cdot \vec{E} = -pE \cos\phi$. This means

$$U_{\phi=0°} = -pE = -2.0 \ \mu\text{J} \qquad U_{\phi=60°} = -pE \cos 60° = -\tfrac{1}{2} pE = -\tfrac{1}{2}(2 \ \mu\text{J}) = -1.0 \ \mu\text{J}$$

The mechanical energy $E_{\text{mech}} = U + K$. We know that at $\phi = 60°$, $K_{\phi=60°} = 0$ J. So,

$$E_{\text{mech}} = U_{\phi=60°} + K_{\phi=60°} = -1.0 \ \mu\text{J} + 0.0 \ \text{J} = -1.0 \ \mu\text{J}$$

(b) Conservation of mechanical energy gives

$$U_{\phi=60°} + K_{\phi=60°} = U_{\phi=0°} + K_{t=0°} \Rightarrow -1.0 \ \mu\text{J} + 0.0 \ \text{J} = -2.0 \ \mu\text{J} + K_{\phi=0°} \Rightarrow K_{\phi=0°} = 1.0 \ \mu\text{J}$$

Section 28.4 The Electric Potential

28.13. Model: Energy is conserved. The potential energy is determined by the electric potential.
Visualize:

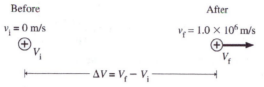

Before   After

$v_i = 0$ m/s $v_f = 1.0 \times 10^6$ m/s

$\Delta V = V_f - V_i$

The figure shows a before-and-after pictorial representation of a He^+ ion moving through a potential difference. The ion's initial speed is zero and its final speed is 2.0×10^6 m/s. A positive charge *speeds up* as it moves into a region of lower potential ($U \rightarrow K$).
Solve: The potential energy of charge q is $U = qV$. Conservation of energy, expressed in terms of the electric potential V, is

$$K_f + qV_f = K_i + qV_i \Rightarrow q(V_f - V_i) = K_i - K_f$$

$$\Rightarrow \Delta V = \frac{K_i - K_f}{q} = \frac{0 \ \text{J} - \frac{1}{2}mv_f^2}{q} = -\frac{4(1.67 \times 10^{-27} \ \text{kg})(2.0 \times 10^6 \ \text{m/s})^2}{2(1.60 \times 10^{-19} \ \text{C})} = -8.4 \times 10^4 \ \text{V}$$

Assess: This result implies that the helium ion moves from a higher potential toward a lower potential.

Section 28.5 The Electric Potential Inside a Parallel-Plate Capacitor

28.17. Model: The electric potential difference between the plates is determined by the uniform electric field in the parallel-plate capacitor.
Solve: (a) The potential of an ordinary AA or AAA battery is 1.5 V. Actually, this is the potential difference between the two terminals of the battery. If the electric potential of the negative terminal is taken to be zero, then the positive terminal is at a potential of 1.5 V.
(b) If a battery with a potential difference of 1.5 V is connected to a parallel-plate capacitor, the potential difference between the two capacitor plates is also 1.5 V. Thus,

$$\Delta V_C = 1.5 \ \text{V} = V_+ - V_- = Ed$$

where d is the separation between the two plates. The electric field inside a parallel-plate capacitor is

$$E = \frac{\eta}{\varepsilon_0} = \frac{Q}{A\varepsilon_0} \Rightarrow 1.5 \ \text{V} = \left(\frac{Q}{A\varepsilon_0}\right)d$$

$$\Rightarrow Q = \frac{(1.5 \ \text{V})(A\varepsilon_0)}{d} = \frac{(1.5 \ \text{V})(4.0 \times 10^{-2} \ \text{m})^2(8.85 \times 10^{-12} \ \text{C}^2/\text{Nm}^2)}{1.0 \times 10^{-3} \ \text{m}} = 2.1 \times 10^{-11} \ \text{C}$$

Thus, the battery moves 2.1×10^{-11} C of electron charge from the positive to the negative plate of the capacitor.

Section 28.6 The Electric Potential of a Point Charge

28.21. Model: The charge is a point charge.
Solve: (a) The electric potential of the point charge q is

$$V = \frac{1}{4\pi\varepsilon_0}\frac{q}{r} = (9.0 \times 10^9 \ \text{N m}^2/\text{C}^2)\left(\frac{2.0 \times 10^{-9} \ \text{C}}{r}\right) = \frac{18.0 \ \text{N m}^2/\text{C}}{r}$$

For points A and B, $r = 0.010$ m. Thus,

$$V_A = V_B = \frac{18.0 \text{ N m}^2/\text{C}}{0.010 \text{ m}} = 1800 \frac{\text{Nm}}{\text{C}} = 1800 \left(\frac{\text{V}}{\text{m}}\right) \text{m} = 1.80 \text{ kV}$$

For point C, $r = 0.020$ m and $V_C = 900$ V.

(b) The potential differences are

$$\Delta V_{AB} = V_B - V_A = 1.80 \text{ kV} - 1.80 \text{ kV} = 0 \text{ V} \qquad \Delta V_{BC} = V_C - V_B = 0.90 \text{ kV} - 1.80 \text{ kV} = -0.90 \text{ kV}$$

28.23. Model: Outside a charged sphere the electric potential is identical to that of a point charge at the center.
Solve: (a) For a proton, assumed to be a point charge, the electric potential is

$$V = \frac{1}{4\pi\varepsilon_0}\frac{(+e)}{r} = (9.0\times10^9 \text{ N m}^2/\text{C}^2)\frac{1.60\times10^{-19} \text{ C}}{0.053\times10^{-9} \text{ m}} = 27 \text{ V}$$

(b) The potential energy of a charge q at a point where the potential is V is $U = qV$. The potential energy of the electron in the proton's potential is

$$U = (-1.60\times10^{-19} \text{ C})(27 \text{ V}) = -4.3\times10^{-18} \text{ J}$$

Section 28.7 The Electric Potential of Many Charges

28.25. Model: The net potential is the sum of the potentials due to each charge.
Solve: From the geometry in the figure,

$$\frac{1.5 \text{ cm}}{r_1} = \frac{1.5 \text{ cm}}{r_2} = \frac{1.5 \text{ cm}}{r_3} = \cos 30° \Rightarrow r_1 = r_2 = r_3 = \frac{1.5 \text{ cm}}{\cos 30°} = 1.732 \text{ cm}$$

The potential at the dot is

$$V = \frac{1}{4\pi\varepsilon_0}\frac{q_1}{r_1} + \frac{1}{4\pi\varepsilon_0}\frac{q_2}{r_2} + \frac{1}{4\pi\varepsilon_0}\frac{q_3}{r_3}$$

$$= (9.0\times10^9 \text{ N m}^2/\text{C}^2)\left[\frac{1.0\times10^{-9} \text{ C}}{0.01732 \text{ m}} - \frac{2.0\times10^{-9} \text{ C}}{0.01732 \text{ m}} - \frac{2.0\times10^{-9} \text{ C}}{0.01732 \text{ m}}\right] = -1600 \text{ V}$$

Assess: Potential is a scalar quantity, so we found the net potential by adding three scalar quantities.

28.29. Model: While the potential is the sum of the scalar potentials due to each charge, the electric field is the vector sum of the electric fields due to each charge.
Solve: (a) As V is always positive, both charges must be positive. The positive signs for the two equal charges at $x = a$ and at $x = b$ are also consistent with the behavior of the potential in the range $a < x < b$.
(b) By the symmetry of the drawing about the middle, we infer that the magnitudes of the charges are the same. Thus $|q_a/q_b| = 1$.
(c)

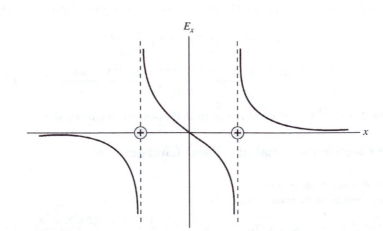

The graph of E_x, the x-component of the electric field, as a function of x is shown in the figure.

28.33. Model: While the net potential is the sum of the potentials due to each charge, the net electric field is the vector sum of the electric fields.
Visualize:

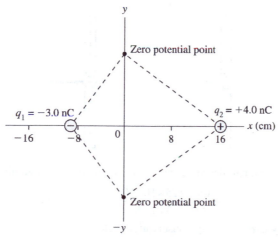

The charge $Q_1 = 20.0$ nC is at the origin. The charge $Q_2 = -10.0$ nC is 15.0 cm to the right of the charge Q_1 on the x-axis.
Solve: (a) As the pictorial representation shows, the point P on the x-axis where the electric field is zero can only be on the right side of the charge Q_2, that is, at $x \geq 15.0$ cm. At this point $E_1 = E_2$, so we have

$$\frac{1}{4\pi\varepsilon_0} \frac{20.0\times10^{-9} \text{ C}}{x^2} = \frac{1}{4\pi\varepsilon_0} \frac{10.0\times10^{-9} \text{ C}}{(x-15.0 \text{ cm})^2} \Rightarrow x^2 = 2(x-15.0 \text{ cm})^2$$

$$\Rightarrow x^2 - (60.0 \text{ cm})x + 450 \text{ cm}^2 = 0 \Rightarrow x = \frac{(60.0 \text{ cm})\pm\sqrt{3600 \text{ cm}^2 -1800 \text{ cm}^2}}{2}$$

$$\Rightarrow x = 51.2 \text{ cm and } 8.8 \text{ cm}$$

The root $x = 8.8$ cm is not possible physically. So, the electric fields cancel out at $x = 51.2$ cm. The electric potential at this point is

$$V = \frac{1}{4\pi\varepsilon_0}\frac{Q_1}{r_1} + \frac{1}{4\pi\varepsilon_0}\frac{Q_2}{r_2} = (9.0\times10^9 \text{ N m}^2/\text{C}^2)\left[\frac{20.0\times10^{-9} \text{ C}}{0.512 \text{ m}} + \frac{-10.0\times10^{-9} \text{ C}}{(0.512 \text{ m}-0.150 \text{ m})}\right] = +103 \text{ V}$$

(b) The point on the x-axis where the potential is zero can be obtained from the condition $V_1 + V_2 = 0$ V, which is

$$\frac{1}{4\pi\varepsilon_0}\frac{Q_1}{r_1} + \frac{1}{4\pi\varepsilon_0}\frac{Q_2}{r_2} = 0 \text{ V} \Rightarrow \frac{20.0\times10^{-9} \text{ C}}{x} + \frac{-10.0\times10^{-9} \text{ C}}{(15.0 \text{ cm} - x)} = 0$$

$$\Rightarrow 2(15.0 \text{ cm} - x) - x = 0 \Rightarrow x = 10.0 \text{ cm}$$

The electric field 10.0 cm away from charge Q_1 is

$$\vec{E}_{net} = \vec{E}_1 + \vec{E}_2 = \frac{1}{4\pi\varepsilon_0}\frac{20.0\times10^{-9} \text{ C}}{(0.100 \text{ m})^2}\hat{i} + \frac{1}{4\pi\varepsilon_0}\frac{(10.0\times10^{-9} \text{ C})}{(0.050 \text{ m})^2}\hat{i} = 5.40\times10^4\hat{i} \text{ V/M}$$

The magnitude is 5.40×10^4 V/M.

28.35. Model: The net potential is the sum of the scalar potentials due to each charge.
Visualize:

Solve: Let the point on the y-axis where the electric potential is zero be at a distance y from the origin. At this point, $V_1 + V_2 = 0$ V. This means

$$\frac{1}{4\pi\varepsilon_0}\left\{\frac{q_1}{r_2} + \frac{q_2}{r_2}\right\} = 0 \text{ V} \Rightarrow \frac{-3.0\times10^{-9} \text{ C}}{\sqrt{(-9.0 \text{ cm})^2 + y^2}} + \frac{4.0\times10^{-9} \text{ C}}{\sqrt{(16.0 \text{ cm})^2 + y^2}} = 0$$

$$\Rightarrow 3\sqrt{(16 \text{ cm})^2 + y^2} = 4\sqrt{(-9 \text{ cm})^2 + y^2} \Rightarrow 9(256 \text{ cm}^2 + y^2) = 16(81 \text{ cm}^2 + y^2)$$

$$\Rightarrow 7y^2 = 1008 \text{ cm}^2 \Rightarrow y = \pm 12 \text{ cm}$$

28.39. Solve:
(a) The work that must be done to move a charged particle through a potential difference is $q\Delta V$.

$$W = q\Delta V = (1.6\times10^{-19} \text{ C})(70 \text{ mV}) = 1.1\times10^{-20} \text{ J}$$

(b) The body uses 20% of 100 J of energy every second to pump sodium ions. We will round the answer to part (a) so we won't need a calculator to get our one-significant-figure answer.

$$\text{number of sodium ions} = \frac{2\times10^1 \text{ J}}{1\times10^{-20} \text{ J/ion}} = 2\times10^{21} \text{ ions}$$

Assess: The answer is a huge number, and even if we divide by the number of cells in our body there are still many sodium ions pumped each second in every cell.

28.41. Model: Energy is conserved.
Solve: The potential energy of the positively charged proton as a function of x is

$$U(x) = eV(x) = 6000ex^2 = 6000(1.60\times10^{-19} \text{ C})x^2 = 9.6\times10^{-16}x^2 \text{ J}.$$

This is analogous to the potential energy of a mass on a spring. Thus the motion is simple harmonic motion.

The form for the potential energy of a mass on a spring is $U(x) = \frac{1}{2}kx^2$. Comparing that to the expression above yields

$$\frac{1}{2}k_{\text{eff}} = 9.6\times10^{-16} \text{ J/m}^2 = 9.6\times10^{-16} \text{ N/m} \Rightarrow k_{\text{eff}} = 1.92\times10^{-15} \text{ N/m}$$

From our chapter on simple harmonic motion we recall

$$f = \frac{1}{2\pi}\sqrt{\frac{k}{m}} = \frac{1}{2\pi}\sqrt{\frac{1.92\times10^{-16} \text{ N/m}}{1.67\times10^{-27} \text{ kg}}} = 54 \text{ kHz}$$

28.43. Model: Energy is conserved.
Visualize:

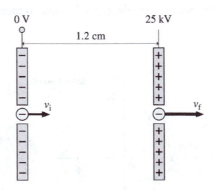

Solve: (a) The electric field inside a parallel-plate capacitor is constant with strength

$$E = \frac{\Delta V}{d} = \frac{(25\times10^3 \text{ V})}{0.012 \text{ m}} = 2.1\times10^6 \text{ V/m}$$

(b) Assuming the initial velocity is zero, energy conservation yields

$$U_i = K_f + U_f$$

$$0 = \frac{1}{2} m_e v_f^2 + (-e)(Ed)$$

$$\Rightarrow v_f = \sqrt{\frac{2(1.60\times10^{-19}\ \text{C})(2.1\times10^6\ \text{V/m})(0.012\ \text{m})}{9.11\times10^{-31}\ \text{kg}}} = 9.4\times10^7\ \text{m/s}$$

Assess: This speed is about 31% the speed of light. At that speed, relativity should be taken into account.

28.47. Model: Energy is conserved. The electron ends up so far away from the glass sphere that we can consider its potential energy to be zero.
Visualize:

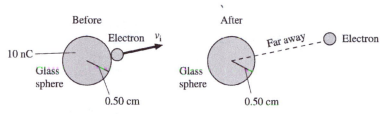

The minimum speed to escape is the speed that allows the electron to reach $r_f = \infty$ when $v_f = 0$ m/s.

Solve: The conservation of energy equation $K_f + U_f = K_i + U_i$ is

$$0\ \text{J} + 0\ \text{J} = \tfrac{1}{2} mv_i^2 + qV_i \Rightarrow 0\ \text{J} = \tfrac{1}{2} mv_i^2 + (-e)\left(\frac{1}{4\pi\varepsilon_0} \frac{q}{R}\right)$$

$$\Rightarrow v_i = \sqrt{\frac{2e}{m} \frac{1}{4\pi\varepsilon_0} \frac{q}{R}} = \sqrt{\frac{2(1.60\times10^{-19}\ \text{C})}{9.11\times10^{-31}\ \text{kg}}(9.0\times10^9\ \text{N m}^2/\text{C}^2)\left(\frac{10\times10^{-9}\ \text{C}}{0.50\times10^{-2}\ \text{m}}\right)} = 8.0\times10^7\ \text{m/s}$$

Assess: This speed is so fast that relativity should be used.

28.49. Model: The electrons and the proton are point charges.
Visualize:

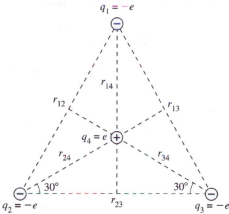

Solve: We are given that $r_{12} = r_{23} = r_{13} = 1.0\times10^{-9}$ m. From the geometry of the figure,

$$\frac{\tfrac{1}{2} r_{23}}{r_{24}} = \cos 30° \Rightarrow r_{24} = \frac{r_{23}}{2\cos 30°} = 0.5774\times10^{-9}\ \text{m} = r_{14} = r_{34}$$

The contributions to the total potential energy are

$$U_{12} = U_{13} = U_{23} = \frac{(9.0\times10^9\ \text{N m}^2/\text{C}^2)(-1.60\times10^{-19}\ \text{C})(-1.60\times10^{-19}\ \text{C})}{1.0\times10^{-9}\ \text{m}} = 2.304\times10^{-19}\ \text{J}$$

$$U_{14} = U_{24} = U_{34} = \frac{(9.0\times10^9\ \text{N m}^2/\text{C}^2)(-1.60\times10^{-19}\ \text{C})(1.60\times10^{-19}\ \text{C})}{0.5774\times10^{-9}\ \text{m}} = -3.990\times10^{-19}\ \text{J}$$

Summing all of the contributions,
$$U_{elec} = U_{12} + U_{13} + U_{23} + U_{14} + U_{24} + U_{34}$$
$$= 3(2.304 \times 10^{-19} \text{ J}) + 3(-3.990 \times 10^{-19} \text{ J}) = -5.1 \times 10^{-19} \text{ J}$$
Assess: Note that $U_{12} = U_{21}$, $U_{13} = U_{31}$, and $U_{23} = U_{32}$, $U_{14} = U_{41}$, $U_{24} = U_{42}$, and $U_{34} = U_{43}$.

28.51. Model: Ignore gravitational effects and consider the motion as one-dimensional in the x-direction. Model the bead as a point charge. As indicated, we assume the bead starts far enough away that the initial electric potential energy is negligible.
Visualize: Because of spherical symmetry we can consider the charge on the Van de Graaff generator to be concentrated at its center.

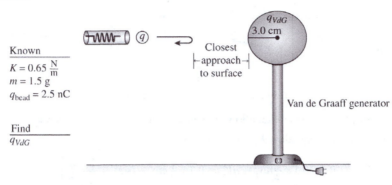

Known
$K = 0.65 \frac{\text{N}}{\text{m}}$
$m = 1.5$ g
$q_{bead} = 2.5$ nC

Find
q_{VdG}

Solve: Use conservation of energy. The spring potential energy before the bead is fired equals the electric potential energy at closest approach (at both times the kinetic energy is zero).
$$U_{sp} = U_{elec}$$
$$\frac{1}{2} k (\Delta x)^2 = K \frac{q_1 q_2}{r}$$
The closest approach r must be measured from the center of the Van de Graaff sphere so we add 3.0 cm to each measurement of the closest approach. Rearrange the equation to produce
$$\frac{1}{r} = \left(\frac{k}{2} \frac{1}{K(q_{bead})(q_{VdG})} \right) (\Delta x)^2$$
This leads us to believe that a graph of $\frac{1}{r}$ vs. $(\Delta x)^2$ would produce a straight line whose slope is $\dfrac{k}{2K(q_{bead})(q_{VdG})}$

and whose intercept is zero.

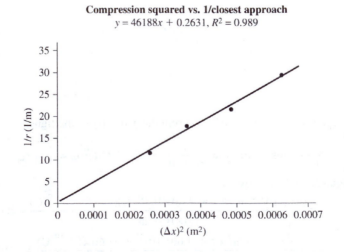

Compression squared vs. 1/closest approach
$y = 46188x + 0.2631, R^2 = 0.989$

We see that the linear fit is good and that the slope is 46188 m^{-3}.

$$q_{VdG} = \frac{k}{2K(q_{bead})(\text{slope})} = \frac{0.65 \text{ N/m}}{2(9.0\times10^9 \text{ N m}^2/\text{C}^2)(2.5 \text{ nC})(46188 \text{ m}^{-3})} = 310 \text{ nC}$$

Assess: The intercept of our graph is very small, as we expect. The units cancel properly. The mass of the bead was irrelevant.

28.57. Model: The electric field inside a capacitor is uniform.
Solve: (a) Because the parallel-plate capacitor was connected to the terminals of a 15 V battery for a long time, the potential difference across the capacitor right after the battery is disconnected is $\Delta V = 15$ V. The electric field strength inside the capacitor is

$$E = \frac{\Delta V_C}{d} = \frac{15 \text{ V}}{0.50\times10^{-2} \text{ m}} = 3000 \text{ V/m} = 3.0 \text{ kV/m}$$

Because $E = \eta/\varepsilon_0$ for a parallel-plate capacitor and $\eta = Q/A$, the total charge on each plate is

$$Q = EA\varepsilon_0 = (3000 \text{ V/m})\pi(0.050 \text{ m})^2(8.85\times10^{-12} \text{ C}^2/\text{Nm}^2) = 2.1\times10^{-10} \text{ C}$$

(b) After the electrodes are pulled away to a separation of $d' = 1.0$ cm, the charges on the plates are unchanged. That is, $Q' = Q$. Because $A' = A$, the electric field inside the capacitor is also unchanged. So, $E' = E$. The potential difference across the capacitor is $\Delta V_C' = E'd'$. Because d increases from 0.50 cm to 1.0 cm ($d' = 1.0$ cm), the potential difference $\Delta V_C'$ increases from 15 V to 30 V.

(c) When the electrodes are expanded, the new area is $A' = \pi(r')^2 = \pi(2r)^2 = 4A$. The charge Q' on the capacitor plates, however, stays the same as before ($Q' = Q$). The electric field is

$$E' = \frac{\eta'}{\varepsilon_0} = \frac{Q'}{A'\varepsilon_0} = \frac{Q}{4A\varepsilon_0} = \frac{E}{4} = \frac{3000 \text{ V/m}}{4} = 750 \text{ V/m} = 0.75 \text{ kV/m}$$

The potential difference across the capacitor plates $\Delta V_C' = E'd' = (750 \text{ V})(0.050 \text{ m}) = 3.8$ V.

28.59. Solve: (a) Outside a uniformly charged sphere, the electric field is identical to that of a point charge Q at the center. For $r \geq R$,

$$E = \frac{1}{4\pi\varepsilon_0}\frac{Q}{r^2}$$

Evaluating this at $r = R$ and using Equation 28.32,

$$E_0 = \frac{1}{4\pi\varepsilon_0}\frac{Q}{R^2} = \frac{V_0}{R}$$

(b) The field strength is

$$E_0 = \frac{500 \text{ V}}{0.50\times10^{-2} \text{ m}} = 100,000 \text{ V/m} = 100 \text{ kV/m}$$

28.63. Model: The potential at any point is the superposition of the potentials due to all charges. Outside a uniformly charged sphere, the electric potential is identical to that of a point charge Q at the center.
Visualize: Sphere A is the sphere on the left and sphere B is the one on the right.
Solve: The potential at point a is the sum of the potentials due to the spheres A and B:

$$V_a = V_{A \text{ at a}} + V_{B \text{ at a}} = \frac{1}{4\pi\varepsilon_0}\frac{Q_A}{R_A} + \frac{1}{4\pi\varepsilon_0}\frac{Q_B}{0.70 \text{ m}}$$

$$= (9.0\times10^9 \text{ N m}^2/\text{C}^2)\frac{100\times10^{-9} \text{ C}}{0.30 \text{ m}} + (9.0\times10^9 \text{ N m}^2/\text{C}^2)\frac{25\times10^{-9} \text{ C}}{0.70 \text{ m}}$$

$$= 3000 \text{ V} + 321 \text{ V} = 3321 \text{ V}$$

Similarly, the potential at point b is the sum of the potentials due to the spheres A and B:

$$V_b = V_{\text{B at b}} + V_{\text{A at b}} = \frac{1}{4\pi\varepsilon_0}\frac{Q_B}{R_B} + \frac{1}{4\pi\varepsilon_0}\frac{Q_A}{0.95\ \text{m}}$$

$$= (9.0\times10^9\ \text{N m}^2/\text{C}^2)\left(\frac{25\times10^{-9}\ \text{C}}{0.05\ \text{m}} + \frac{100\times10^{-9}\ \text{C}}{0.95\ \text{m}}\right)$$

$$= 4500\ \text{V} + 947\ \text{V} = 5447\ \text{V}$$

Thus, the potential at point b is higher than the potential at a. The difference in potential is $V_b - V_a = 5447\ \text{V} - 3321\ \text{V} = 2126\ \text{V} = 2.1\ \text{kV}.$

Assess: $V_{\text{A at a}} = 3000\ \text{V}$ and the sphere B has a potential of 225 V at point a. The spherical symmetry dictates that the potential on a sphere's surface be the same everywhere. So, in calculating the potential at point a due to the sphere B we used the center-to-center separation of 1.0 m rather than a separation of $100\ \text{cm} - 30\ \text{cm} = 70\ \text{cm}$ from the center of sphere B to the point a. The former choice leads to the same potential everywhere on the surface whereas the latter choice will lead to a distribution of potentials depending upon the location of point a. Similar reasoning also applies to the potential at point b.

28.65. Model: The potential is the sum of potentials from each charge.
Visualize:

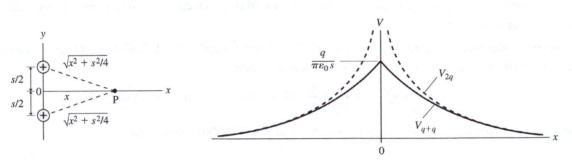

Point P is located on the x-axis and the two positive charges are located on the y-axis. The separation between the two charges is s.
Solve: (a) The potentials at P due to the two positive charges are

$$V_{\text{top}} = \frac{1}{4\pi\varepsilon_0}\frac{(+q)}{\sqrt{x^2 + s^2/4}} \qquad V_{\text{bottom}} = \frac{1}{4\pi\varepsilon_0}\frac{(+q)}{\sqrt{x^2 + s^2/4}}$$

$$\Rightarrow V_{q+q} = V_{\text{top}} + V_{\text{bottom}} = \frac{1}{4\pi\varepsilon_0}\frac{2q}{\sqrt{x^2 + s^2/4}} = \frac{2q}{4\pi\varepsilon_0 x}\frac{1}{\sqrt{1 + s^2/4x^2}}$$

(b) The V-versus-x graph is shown in the figure. The potential due to a charge $2q$ located at the origin is $V_{2q} = 2q/4\pi\varepsilon_0 x$, and is also plotted. This expression differs from the expression obtained in part (a) by the factor $(1 + s^2/4x^2)^{-\frac{1}{2}}$. For $x \gg s$, this factor becomes 1 and the two potentials become the same.

28.67. Model: Assume the thin rod is a line of charge with uniform linear charge density.
Visualize: The point P is a distance d from the origin. Divide the charged rod into N small segments, each of length Δx and with charge Δq. Segment i, located at position x_i, contributes a small amount of potential V_i at point P.
Solve: The contribution of the ith segment is

$$V_i = \frac{\Delta q}{4\pi\varepsilon_0 r_i} = \frac{\Delta q}{4\pi\varepsilon_0 (d - x_i)} = \frac{Q\Delta x/L}{4\pi\varepsilon_0 (d - x_i)}$$

where $\Delta q = \lambda\Delta x$ and the linear charge density is $\lambda = Q/L$. We are placing the point P at a distance d rather than x from the origin to avoid confusion with x_i. The V_i are now summed and the sum is converted to an integral giving

$$V = \frac{Q}{4\pi\varepsilon_0 L}\int_{-L/2}^{L/2}\frac{dx}{d-x} = \frac{Q}{4\pi\varepsilon_0 L}\left[-\ln(d-x)\right]_{-L/2}^{L/2} = \frac{Q}{4\pi\varepsilon_0 L}\ln\left(\frac{d+L/2}{d-L/2}\right)$$

Replacing d with x, the potential due to a line charge of length L at a distance x along the axis is

$$V = \frac{Q}{4\pi\varepsilon_0 L}\ln\left(\frac{x+L/2}{x-L/2}\right)$$

28.69. Model: Because the rod is thin, assume the charge lies along the semicircle of radius R.
Visualize: The bent rod lies in the xy-plane with point P as the center of the semicircle. Divide the semicircle into N small segments of length Δs and of charge $\Delta Q = (Q/\pi R)\Delta s$, each of which can be modeled as a point charge. The potential V at P is the sum of the potentials due to each segment of charge.
Solve: The total potential is

$$V = \sum_i V_i = \sum_i \frac{1}{4\pi\varepsilon_0}\frac{\Delta Q}{R} = \frac{1}{4\pi\varepsilon_0 R}\sum_i \left(\frac{Q}{\pi R}\right)\Delta s = \frac{1}{4\pi\varepsilon_0}\frac{Q}{\pi R^2}\sum_i R\Delta\theta = \frac{1}{4\pi\varepsilon_0}\frac{Q}{\pi R}\sum_i \Delta\theta.$$

All of the terms come to the front of the summation because these quantities did not change as far as the summation is concerned. The summation does not have to convert to an integral because the sum of all the $\Delta\theta$ around the semicircle is π. Hence, the potential at the center of a charged semicircle is

$$V_{\text{center}} = \frac{1}{4\pi\varepsilon_0}\frac{Q\pi}{\pi R} = \frac{1}{4\pi\varepsilon_0}\frac{Q}{R}$$

POTENTIAL AND FIELD

Exercises and Problems

Section 29.1 Connecting Potential and Field

29.3. Model: The potential difference is the negative of the area under the E_x vs. x curve.
Visualize: Please refer to Figure EX29.3.
Solve: The potential difference between $x = 1.0$ m and $x = 3.0$ m is

$$\Delta V = -\frac{1}{2}(200 \text{ V})(3.0 \text{ m} - 1.0 \text{ m}) = -200 \text{ V}$$

Assess: The potential difference is negative since the electric field points in the direction of decreasing potential.

Section 29.2 Sources of Electric Potential

29.5. Solve: The work done is exactly equal to the increase in the potential energy of the charge. That is,

$$W = \Delta U = q\Delta V = q(V_f - V_i) = (1.0 \times 10^{-6} \text{ C})(1.5 \text{ V}) = 1.5 \times 10^{-6} \text{ J}$$

Assess: The work done by the escalator on the charge is stored as electric potential energy of the charge.

29.7. Solve: The work done is the change in potential energy of the charge, and the electric potential is the potential energy per unit charge, so

$$W = \Delta U = q\Delta V \quad \Rightarrow \quad q = \frac{W}{\Delta V} = \frac{27 \text{ J}}{9.0 \text{ V}} = 3.0 \text{ C}$$

Section 29.3 Finding the Electric Field from the Potential

29.9. Model: The electric field points in the direction of *decreasing* potential and is perpendicular to the equipotential lines.
Visualize: Please refer to Figure EX29.9. The three equipotential surfaces correspond to potentials of 0 V, 200 V, and 400 V.
Solve: The electric field component along a direction of constant potential is $E_s = -dV/ds = 0$ V/m. But, the electric field component perpendicular to the equipotential surface is

$$\left|\vec{E}\right| = \frac{\Delta V}{\Delta s} = \frac{400 \text{ V}}{0.02 \text{ m}} = 20 \text{ kV/m}$$

The direction of the electric field vector is "downhill," perpendicular to the equipotential surfaces. That is, the electric field is 20 kV/m downward or $\vec{E} = -(20\hat{j})$ kV/m.

29.15. Solve: (a) Since $E_x = -dV/dx$, we have

$$E_x = -\frac{d}{dx}(100e^{-2x}) \text{ V/m} = 200e^{-2x} \text{ V/m}$$

At $x = 1.0$ m,

$$E_x = (200e^{-2(1.0 \text{ m})}) \text{ V/m} = 27 \text{ V/m}$$

(b) At $x = 2.0$ m,

$$E_x = (200e^{-2(2.0 \text{ m})}) \text{ V/m} = 3.7 \text{ V/m}$$

Section 29.5 Capacitance and Capacitors

29.17. Model: Assume that the battery is ideal and that the capacitor is a parallel-plate capacitor.
Solve: (a) From Equation 29.17, the capacitance is

$$C = \frac{\epsilon_0 A}{d} = \frac{(8.85\times10^{-12} \text{ C}^2/\text{N m}^2)\pi(0.015 \text{ m})^2}{0.00050 \text{ m}} = 1.25\times10^{-11} \text{ F} = 13 \text{ pF}$$

(b) The magnitude of the charge on each electrode is

$$Q = C\Delta V_C = (1.25\times10^{-12} \text{ F})(100 \text{ V}) = 1.25\times10^{-10} \text{ C} = 0.13 \text{ nC}$$

29.21. Solve: According to Equation 29.21,

$$C_{eq} = C_1 + C_2 + C_3 = 6 \text{ } \mu F + 10 \text{ } \mu F + 16 \text{ } \mu F = 32 \text{ } \mu F$$

29.23. Model: Two capacitors in series combine to give less capacitance.
Solve: Since we have a 75 μF capacitor and we want a 50 μF capacitance, we must connect the second capacitor in series with the 75 μF capacitor. The capacitance of the second capacitor is calculated as follows:

$$\frac{1}{75 \text{ } \mu F} + \frac{1}{C} = \frac{1}{50 \text{ } \mu F} \quad \Rightarrow \quad C = 150 \text{ } \mu F$$

Section 29.6 The Energy Stored in a Capacitor

29.27. Solve: Using Equation 29.26,

$$U_{C1} = \tfrac{1}{2}C_1(\Delta V_{C1})^2 \qquad U_{C2} = \tfrac{1}{2}C_2(\Delta V_{C2})^2$$

Because $C_2 = \tfrac{1}{2}C_1$ and $\Delta V_{C2} = 2\Delta V_{C1}$, we have

$$\frac{U_{C1}}{U_{C2}} = \frac{\tfrac{1}{2}C_1(\Delta V_{C1})^2}{\tfrac{1}{2}C_2(\Delta V_{C2})^2} = \frac{C_1(\Delta V_{C1})^2}{\left(\tfrac{1}{2}C_1\right)4(\Delta V_{C1})^2} = \frac{1}{2}$$

29.29. Solve: (a) From Equation 29.26, the energy stored in the charged capacitor is

$$U_C = \frac{1}{2}C(\Delta V_C)^2 = \frac{1}{2}\left(\frac{A\epsilon_0}{d}\right)(\Delta V_C)^2$$

$$= \frac{1}{2}\frac{\pi(0.010 \text{ m})^2(8.85\times10^{-12} \text{ C}^2/\text{N m}^2)}{0.50\times10^{-3} \text{ m}}(200 \text{ V})^2 = 1.1\times10^{-7} \text{ J}$$

(b) From Equation 29.28, the energy density in the electric field is

$$u_E = \frac{1}{2}\epsilon_0 E^2 = \frac{1}{2}\epsilon_0\left(\frac{\Delta V_C}{d}\right)^2 = \frac{1}{2}(8.85\times10^{-12} \text{ C}^2/\text{N m}^2)\left(\frac{200 \text{ V}}{0.50\times10^{-3} \text{ m}}\right)^2 = 0.71 \text{ J/m}^3$$

Section 29.7 Dielectrics

29.31. Model: The electrodes form a parallel-plate capacitor.
Solve: (a) The capacitance of the equivalent vacuum-filled capacitor is

$$C_0 = \frac{\epsilon_0 A}{d} = \frac{(8.85\times10^{-12} \, C^2/Nm^2)(5.0\times10^{-3} \, m)^2}{(0.10\times10^{-3} \, m)} = 2.21\times10^{-12} \, F$$

From Table 29.1, the dielectric constant of Mylar is $\kappa = 3.1$. With the battery attached, the potential difference across the plates is $\Delta V_C = \Delta V_{batt} = 9.0$ V. The charge on the plates is

$$Q = C\Delta V = \kappa C_0 \Delta V = (3.1)(2.21\times10^{-12} \, F)(9.0 \, V) = 6.17\times10^{-11} \, C = 62 \text{ pC}$$

The electric field inside the capacitor is

$$E = \frac{E_0}{\kappa} = \frac{1}{\kappa}\frac{\Delta V_C}{d} = \left(\frac{1}{3.1}\right)\left(\frac{9.0 \, V}{0.10\times10^{-3} \, m}\right) = 29 \text{ kV/m}$$

(b) With the battery connected, $\Delta V_C = 9.0$ V. The capacitor is now vacuum-insulated.

$$E = E_0 = \frac{\Delta V_C}{d} = \frac{9.0 \, V}{0.10\times10^{-3} \, m} = 90 \text{ kV/m}$$

$$Q = C_0 \Delta V_C = (2.21\times10^{-12} \, F)(9.0 \, V) = 20 \text{ pC}$$

Assess: Since the battery remains connected as the Mylar is withdrawn, the potential difference across the plates does not change.

29.35. Solve: (a) See plots below.

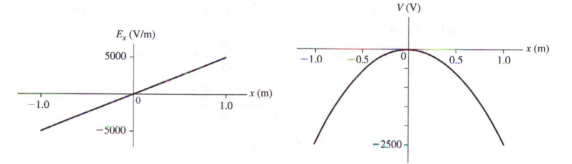

(b) Equation 29.3 gives the potential difference between two points in space:

$$\Delta V = V(x_f) - V(x_i) = -\int_{x_i}^{x_f} E_x \, dx = -\int_{x_i}^{x_f} (5000x \, V/m) \, dx = -5000\left[\frac{x^2}{2}\right]_{x_i}^{x_f} V = -2500(x_f^2 - x_i^2) \, V$$

Taking $V(x_i) = 0$ V at $x_i = 0$ m, and replacing x_f with simply x gives $V(x) = -(2500 \, x^2)$ V.
(c) A graph of V versus x over the region $-1 \, m \le x \le 1 \, m$ is shown in part (a).
Assess: As it must be, we have

$$-\frac{dV}{dx} = -\frac{d}{dx}(-2500x^2) \, V = 5000x = E_x$$

29.39. Model: The electric field is the negative of the slope of the potential function.
Solve: The on-axis potential of a charged disk obtained in Chapter 28 was

$$V_{disk} = \frac{Q}{2\pi\epsilon_0 R^2}\left[\sqrt{R^2 + z^2} - z\right]$$

where the charge on the disk of radius R is Q and the point is a distance z away from the center of the disk. Because the magnitude E of the electric field is $E = -dV/dz$, we have

$$E_{disk} = -\frac{dV_{disk}}{dz} = -\frac{Q}{2\pi\epsilon_0 R^2}\left[\frac{\frac{1}{2}(2z)}{\sqrt{R^2+z^2}} - 1\right] = \frac{Q}{2\pi\epsilon_0 R^2}\left[1 - \frac{z}{\sqrt{R^2+z^2}}\right]$$

If $Q > 0$, the electric field lines will point away from the disk, so the electric field is

$$\vec{E}_{disk}(z) = \frac{Q}{2\pi\epsilon_0 R^2}\left[1 - \frac{z}{\sqrt{R^2+z^2}}\right]\hat{k}$$

Assess: Using binomial expansion with $z \gg R$ we have

$$E_{disk} = \frac{Q}{2\pi\epsilon_0 R^2}\left[1 - \frac{z}{z(1+R^2/z^2)^{1/2}}\right] = \frac{Q}{2\pi\epsilon_0 R^2}\left[1 - (1+R^2/z^2)^{-1/2}\right] = \frac{Q}{2\pi\epsilon_0 R^2}\left[1 - 1 + \frac{1}{2}R^2/z^2 + \cdots\right] = \frac{Q}{4\pi\epsilon_0 z^2}$$

That is, the disk behaves like a point charge. This is an expected result.

29.41. Model: The electric field is the negative of the slope of the potential graph.
Visualize: Please refer to Figure P29.41.
Solve: Since the contours are uniformly spaced along the y-axis above and below the origin, the slope method is the easiest to apply. Point 1 is in the center of a 75 V change (25 V to 100 V) over a distance of 2 cm, so the slope $\Delta V/\Delta s$ is 37.5 V/cm or 3750 V/m. Point 2 has the same potential difference in half the distance. Thus the slope at point 2 is 7500 V/m. Thus, the magnitudes of the electric fields at points 1 and 2 are 3750 V/m and 7500 V/m. The directions of the electric fields are downward at point 1 and upward at point 2; that is, from the higher potential to the lower potential. All together, this gives
$$\vec{E}_1 = (3750 \text{ V/m, down}) \qquad \vec{E}_2 = (7500 \text{ V/m, up})$$

29.43. Model: The electric field is the negative of the slope of the graph of the potential function.
Solve: The electric potential in a region of space is $V = (150x^2 - 200y^2)$ V where x and y are in meters. The x- and y-components of the electric field are
$$E_x = -\frac{dV}{dx} = -(300\,x) \text{ V/m} \qquad E_y = -\frac{dV}{dy} = +(400\,y) \text{ V/m}$$

At $(x,y) = (2.0 \text{ m}, 2.0 \text{ m})$, $E_x = -600$ V/m and $E_y = 800$ V/m. The magnitude and direction of the electric field are
$$E = \sqrt{E_x^2 + E_y^2} = \sqrt{(-600 \text{ V/m})^2 + (800 \text{ V/m})^2} = 1000 \text{ V/m}$$
$$\tan\theta = \frac{E_y}{|E_x|} = \frac{800 \text{ V/m}}{600 \text{ V/m}} = \frac{4}{3} \quad \Rightarrow \quad \theta = 53° \text{ above the } -x\text{-axis}$$

The electric field points $180° - 53° = 127°$ counterclockwise from the $+x$-axis.

29.47. Model: The potential inside a conductor is the same as the potential on the surface.
Solve: The potential on the surface of the copper sphere must be 500 V. That is,
$$500 \text{ V} = \frac{1}{4\pi\epsilon_0}\frac{Q}{R} \quad \Rightarrow \quad Q = (4\pi\epsilon_0)R(500 \text{ V}) \quad \Rightarrow \quad = \frac{(0.020 \text{ m})(500 \text{ V})}{9.0\times10^9 \text{ N m}^2/\text{C}} = 1.1 \text{ nC}$$

29.49. Model: Assume the battery is ideal.
Visualize:

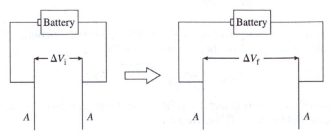

The pictorial representation shows the capacitor plates connected to a battery, and the capacitor plates moved apart with insulating handles while the plates are connected to the battery.

Solve: (a) The initial capacitance of the plates is

$$C_i = \frac{\epsilon_0 A}{d_i} = \frac{(8.85 \times 10^{-12} \text{ C}^2/\text{N m}^2)(0.020 \text{ m})^2}{1.0 \times 10^{-3} \text{ m}} = 3.54 \times 10^{-12} \text{ F}$$

Consequently, an initial voltage $\Delta V_i = 9.0$ V charges the plates to

$$Q = \pm C_i \Delta V_i = \pm(3.54 \times 10^{-12} \text{ F})(9.0 \text{ V}) = \pm 31.9 \times 10^{-12} \text{ C} = \pm 32 \text{ pC}$$

(b) The new capacitance is $C_f = \epsilon_0 A/2d_i = \frac{1}{2} C_i$. The potential difference across the plates is determined by the battery and is unchanged: $\Delta V_f = \Delta V_i = 9.0$. Thus, the new charge on the plates is

$$Q = \pm C_i \Delta V_f = \pm \frac{1}{2}(3.54 \times 10^{-12} \text{ F})(9.0 \text{ V}) = \pm 16.0 \times 10^{-12} \text{ C} = \pm 16.0 \text{ pC}$$

29.53. Model: Assume that the battery is ideal.

Visualize:

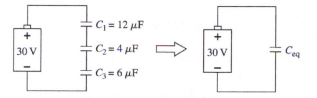

The pictorial representation shows the equivalent capacitance of the three capacitors.

Solve: Because C_1, C_2, and C_3 are in series,

$$\frac{1}{C_{eq}} = \frac{1}{C_1} + \frac{1}{C_2} + \frac{1}{C_3} = \frac{1}{12 \ \mu F} + \frac{1}{4 \ \mu F} + \frac{1}{6 \ \mu F} = \frac{1}{2}(\mu F)^{-1} \quad \Rightarrow \quad C_{eq} = 2 \ \mu F$$

A potential difference of $\Delta V_C = \varepsilon = 30$ V across a capacitor of equivalent capacitance $2 \ \mu F$ produces a charge $Q = C_{eq} \Delta V_C = (2\mu F)(30 \text{ V}) = 60 \mu C$. Because C_{eq} is a combination of three series capacitors, $Q_1 = Q_2 = Q_3 = 60 \mu C$. We are now able to find the potential difference across each capacitor:

$$\Delta V_1 = \frac{Q_1}{C_1} = \frac{60 \ \mu C}{12 \ \mu F} = 5.0 \ V \qquad \Delta V_2 = \frac{Q_2}{C_2} = \frac{60 \ \mu C}{4 \ \mu F} = 15 \ V \qquad \Delta V_3 = \frac{Q_3}{C_3} = \frac{60 \ \mu C}{6 \ \mu F} = 10 \ V.$$

Assess: $\Delta V_1 + \Delta V_2 + \Delta V_3 = 30 \text{ V} = \Delta V_{bat}$, as it should.

29.57. Model: Assume the battery is ideal.

Visualize:

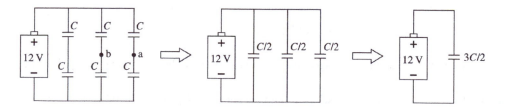

The circuit in Figure P29.57 has been redrawn to show that the six capacitors are arranged in three parallel combinations, each combination being a series combination of two capacitors.

Solve: (a) The equivalent capacitance of the two capacitors in series is $\frac{1}{2}C$. The equivalent capacitance of the six capacitors is $\frac{3}{2}C$.

(b) As points a and b are midpoints between identical capacitors, $V_a = V_b = 6.0$ V. Therefore, the potential difference between points a and b is zero.

29.59. Model: Assume the battery is ideal.
Visualize: Please refer to Figure P29.59. While the switch is in position A, the capacitors C_2 and C_3 are uncharged. When the switch is placed in position B, the charged capacitor C_1 is connected to C_2 and C_3. C_2 and C_3 are connected in series to form an equivalent capacitor $C_{eq\ 23}$.
Solve: While the switch is in position A, a potential difference of $V_1 = 100$ V across C_1 charges it to

$$Q_1 = C_1 V_1 = (15\ \mu F)(100\ V) = 1500\ \mu C$$

When the switch is moved to position B, this initial charge Q_1 is redistributed. The charge Q_1' goes on C_1 and the charge $Q_{eq\ 23}$ goes on $C_{eq\ 23}$. The voltage across C_1 and $C_{eq\ 23}$ is the same and $Q_1' + Q_{eq\ 23} = Q_1 = 1500\ \mu C$. Combining these two conditions, we get

$$\frac{Q_1'}{C_1} = \frac{Q_{eq\ 23}}{C_{eq\ 23}} \quad \Rightarrow \quad \frac{1500\ \mu C - Q_{eq\ 23}}{C_1} = \frac{Q_{eq\ 23}}{C_{eq\ 23}}$$

Since $C_{eq\ 23} = \left(\frac{1}{20\ \mu F} + \frac{1}{30\ \mu F}\right)^{-1} = 12\ \mu F$, we can rewrite this equation as

$$\frac{1500\ \mu C - Q_{eq\ 23}}{15\ \mu F} = \frac{Q_{eq\ 23}}{12\ \mu F} \quad \Rightarrow \quad Q_{eq\ 23} = 0.67\ mC \quad \Rightarrow \quad Q_1' = Q_1 - Q_{eq\ 23} = 1.500\ mC - 0.67\ mC = 0.83\ mC$$

Having found the charge $Q_{eq\ 23}$, it is easy to see that $Q_2 = Q_3 = 0.67\ mC$ because $C_{eq\ 23}$ is a series combination of C_2 and C_3. Thus,

$$\Delta V_1 = \frac{Q_1'}{C_1} = \frac{830\ \mu C}{15\ \mu F} = 55\ V \qquad \Delta V_2 = \frac{Q_2}{C_2} = \frac{670\ \mu C}{20\ \mu F} = 34\ V \qquad \Delta V_3 = \frac{Q_3}{C_3} = \frac{670\ \mu C}{30\ \mu F} = 22\ V$$

29.63. Solve: (a) The capacitance of the parallel-plate capacitor is

$$C_1 = \frac{A\epsilon_0}{d_1} = \frac{(0.10\ m)^2 (8.85 \times 10^{-12}\ C^2/N\ m^2)}{1.0 \times 10^{-3}\ m} = 8.85 \times 10^{-11}\ F$$

The electric potential energy stored in the capacitor is

$$U_1 = \frac{1}{2} \frac{Q^2}{C_1} = \frac{1}{2} \frac{(10 \times 10^{-9}\ C)^2}{8.85 \times 10^{-11}\ F} = 5.7 \times 10^{-7}\ J$$

(b) There is no change in the charge. The energy change is due to the change in the capacitance. The new capacitance is

$$C_2 = \frac{\epsilon_0 A}{d_2} = \frac{\epsilon_0 A}{2d_1} = \frac{C_1}{2}$$

The amount of energy stored is

$$U_2 = \frac{1}{2} \frac{Q^2}{C_2} = 2U_1 = 11.7 \times 10^{-7}\ J$$

(c) Work was done on the capacitor by the agent pulling the plates apart, thereby adding energy into the system.

29.65. Solve: The energy total energy discharged into the patient is $U_C = Pt = (6500\ W)(5.0 \times 10^3\ s) = 32.5$ J. This energy was stored in the capacitor and may be found from Equation 29.27:

$$U_C = \frac{1}{2} C(\Delta V_C)^2 \quad \Rightarrow \quad \Delta V_C = \pm\sqrt{\frac{2U_C}{C}} = \pm\sqrt{\frac{2(32.5\ J)}{90\ \mu F}} = \pm 0.85\ kV$$

The two signs indicate that we do not know the polarity of the capacitor. However, we can say that the capacitor is charged to 0.85 kV.

29.67. Solve: The work needed to lift the copper block is equal to its change in potential energy, or $W = mgh$. This work has to be supplied by the 90% of the motor's output, which comes from the energy stored in the capacitor. Thus,

$$U_C(0.90) = W \quad \Rightarrow \quad \frac{1}{2}C\Delta V_C^2(0.90) = mgh \quad \Rightarrow \quad C = \frac{2mgh}{\Delta V_C^2(0.90)} = \frac{2(2.0\text{ kg})(9.8\text{ m/s}^2)(3.0\text{ m})}{(1000\text{ V})(0.90)} = 0.13\text{ F}$$

29.69. Model: Model the cell membrane as a parallel-plate capacitor and the cell as a spheroid.

Solve: The energy density stored in the electric field in the cell membrane is $u_E = \epsilon_0 E^2/2$ and the electric field is $E = \Delta V/d$, where $\Delta V = -70$ mV is the potential difference across the membrane and $d = 7.0$ nm is the membrane thickness. Combining these equations gives

$$u_E = \frac{\epsilon_0}{2}\left(\frac{\Delta V}{d}\right)^2$$

The total energy stored in the cell membrane is its volume times the energy density, or

$$U_C = (V_{\text{out}} - V_{\text{in}})u_E = \frac{4\pi r^3}{3}\left[1 - \left(1 - \frac{d}{r}\right)^3\right]\frac{\epsilon_0}{2}\left(\frac{\Delta V}{d}\right)^2$$

$$= \frac{4\pi(25\times10^{-6}\text{ m})^3}{3}\left[1 - \left(1 - \frac{7.0\times10^{-9}\text{ m}}{25\times10^{-6}\text{ m})}\right)^3\right]\frac{8.85\times10^{-12}\text{ C}^2/\text{Nm}^2}{2}\left(\frac{-70\times10^{-3}\text{ V}}{7.0\times10^{-9}\text{ m}}\right)^2 = 2.4\times10^{-14}\text{ J}$$

CURRENT AND RESISTANCE

Exercises and Problems

Section 30.1 The Electron Current

30.3. Solve: The number of electrons crossing a cross-sectional area of a wire is the electron current i.

$$i = nAv_d = (6.0 \times 10^{28} \text{ m}^{-3})\pi \left(\frac{1.6 \times 10^{-3} \text{ m}}{2}\right)^2 (2.0 \times 10^{-4} \text{ m/s}) = 2.4 \times 10^{19} \text{ s}^{-1}$$

The electron density n for aluminum is taken from Table 30.1. The number of electrons passing through the cross section in one day is

$$N_e = i\Delta t = (2.4 \times 10^{19} \text{ s}^{-1})(365 \text{ days})(24 \text{ hr/day})(3600 \text{ s/hr}) = 7.6 \times 10^{26} \text{ electrons}$$

Assess: The large electron density compensates for the small drift velocity to deliver a huge number of electrons in current.

Section 30.2 Creating a Current

30.5. Model: Use the conduction model to relate the drift speed to the electric field strength.
Solve: From Equation 30.7, the electric field is

$$E = \frac{mv_d}{e\tau} = \frac{(9.11 \times 10^{-31} \text{ kg})(2.0 \times 10^{-4} \text{ m/s})}{(1.60 \times 10^{-19} \text{ C})(5.0 \times 10^{-14} \text{ s})} = 0.023 \text{ V/m}$$

30.7. Model: We will use the model of conduction to relate the electric field strength to the mean free time between collisions.
Solve: From Equation 30.8, the electric field is

$$i = \frac{ne\tau A}{m} E = \frac{(8.5 \times 10^{28} \text{ m}^{-3})(1.6 \times 10^{-19} \text{ C})(4.2 \times 10^{-15} \text{ s})\pi(0.9 \times 10^{-3} \text{ m})^2}{(9.11 \times 10^{-31} \text{ kg})} 0.065 \text{ V/m} = 1.6 \times 10^{20} \text{ s}^{-1}$$

Section 30.3 Current and Current Density

30.11. Solve: From Equation 30.13, the current in the wire is

$$I = JA = (7.50 \times 10^5 \text{ A/m}^2)(2.5 \times 10^{-6} \text{ m} \times 75 \times 10^{-6} \text{ m}) = 0.141 \text{ mA}$$

The charge that flows in 15 min is the current times the time.

$$Q = I\Delta t = (0.141 \text{ A})(900 \text{ s}) = 127 \text{ C} \approx 130 \text{ C}$$

30.13. Solve: Equation 30.10 is $Q = I\Delta t$. The amount of charge delivered is

$$Q = (10.0 \text{ A})\left(5.0 \text{ min} \times \frac{60 \text{ s}}{1 \text{ min}}\right) = 3.0 \times 10^3 \text{ C}$$

The number of electrons that flow through the hair dryer is

$$N = \frac{Q}{e} = \frac{3.0 \times 10^3 \text{ C}}{1.60 \times 10^{-19} \text{ C}} = 1.88 \times 10^{22}$$

30.15. Visualize:

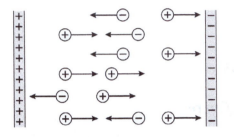

The direction of the current $\vec{I}$ in a material is opposite to the direction of motion of the negative charges and is the same as the direction of motion of positive charges.
Solve: The charge due to positive ions moving to the right per second is

$$q_+ = N_+(2e) = (5.0 \times 10^{15})(2 \times 1.60 \times 10^{-19} \text{ C}) = 1.60 \times 10^{-3} \text{ C}$$

The charge due to negative ions moving to the left per second is

$$q_- = N(-e) = (6.0 \times 10^{15} \text{ s})(-1.60 \times 10^{-19} \text{ C}) = -0.96 \times 10^{-3} \text{ C}$$

Thus, the current in the solution is

$$i = \frac{q_+ - q_-}{t} = \frac{1.60 \times 10^{-3} \text{ C} - (-0.96 \times 10^{-3} \text{ C})}{1 \text{ s}} = 2.56 \times 10^{-3} \text{ A} = 2.6 \text{ mA}$$

Section 30.4 Conductivity and Resistivity

30.19. Solve: The current density is $J = \sigma E$. Using Equation 30.17 and Table 30.2, the current in the wire is

$$I = \sigma E A = (3.5 \times 10^7 \text{ } \Omega^{-1}\text{m}^{-1})(0.012 \text{ V/m})(4 \times 10^{-6} \text{ m}^2) = 1.68 \text{ A}$$

30.23. Solve: (a) Since $J = \sigma E$ and $J = I/A$, the electric field is

$$E = \frac{I}{\sigma A} = \frac{I}{\pi r^2 \sigma} = \frac{0.020 \text{ A}}{\pi (0.25 \times 10^{-3} \text{ m})^2 (6.2 \times 10^7 \text{ } \Omega^{-1}\text{m}^{-1})} = 1.64 \times 10^{-3} \text{ V/m}$$

(b) Since the current density is related to v_d by $J = I/A = nev_d$, the drift speed is

$$v_{\text{d}} = \frac{I}{\pi r^2 ne} = \frac{0.020 \text{ A}}{\pi (0.25 \times 10^{-3} \text{ m})^2 (5.8 \times 10^{28} \text{ m}^{-3})(1.60 \times 10^{-19} \text{ C})} = 1.10 \times 10^{-5} \text{ m/s}$$

Assess: The values of n and σ for silver have been taken from Table 30.1 and Table 30.2. The drift velocity is typical of metals.

30.25. Visualize:

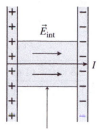

1 cm × 1 cm × 1 cm block

Solve: The current density through the cube is $J = \sigma E$, and the actual current is $I = AJ$. Combining these equations, the conductivity is

$$\sigma = \frac{I}{AE} = \frac{9.0 \text{ A}}{(10^{-4} \text{ m}^2)(5.0 \times 10^{-3} \text{ V/m})} = 1.80 \times 10^7 \; \Omega^{-1} \text{m}^{-1}$$

If *all* quantities entering the calculation are in SI units then the result for σ has to be in SI units. From the value for σ, we can identify the metal as being tungsten.

30.5 Resistance and Ohm's Law

30.29. Solve: We can identify the material by its resistivity. Starting with the wire's resistance,

$$R = \frac{\rho L}{A} = \frac{\rho L}{\pi r^2} \Rightarrow \rho = \frac{\pi r^2 R}{L} = \frac{\pi (0.0004 \text{ m})^2 (1.1 \; \Omega)}{10 \text{ m}} = 5.5 \times 10^{-8} \; \Omega\text{m}$$

From Table 30.2, we can identify the wire as being made of tungsten.

30.31. Model: Assume the battery is an ideal battery.
Solve: (a) The current in a wire is related to its resistance R and to the potential difference ΔV_{wire} between the ends of the wire as

$$I = \frac{\Delta V_{\text{wire}}}{R} \Rightarrow R = \frac{\Delta V_{\text{wire}}}{I} = \frac{1.5 \text{ V}}{0.50 \text{ A}} = 3.0 \; \Omega$$

From Equation 30.22,

$$R = \frac{\rho L}{A} \Rightarrow L = \frac{RA}{\rho} = \frac{(3.0 \; \Omega)\pi(0.30 \times 10^{-3} \text{ m})^2}{2.8 \times 10^{-8} \; \Omega\text{m}} = 30 \text{ m}$$

(b) For a uniform wire of the same material, $R \propto L$. If the wire in this problem is cut in half, its resistance will decrease to $\frac{1}{2}(3\,\Omega) = 1.5 \; \Omega$. Thus, using $I = \Delta V_{\text{wire}}/R$, we have $I = 1.0$ A.

30.35. Solve: From Table 30.2, the resistivity of aluminum is $\rho = 2.8 \times 10^{-8} \; \Omega$ m. From Equation 30.22, the length L of a wire with a cross-sectional area A and having a resistance R is

$$L = \frac{AR}{\rho} = \frac{(10 \times 10^{-6} \text{ m})^2 (1000 \; \Omega)}{2.8 \times 10^{-8} \; \Omega \text{ m}} = 3.57 \text{ m}$$

The number of turns is the length of the wire divided by the circumference of one turn. Thus,

$$\frac{3.57 \text{ m}}{2\pi(1.5 \times 10^{-3} \text{ m})} = 380$$

30.37. Solve: We need an aluminum wire whose resistance and length are the same as that of a 0.50-mm-diameter copper wire. That is,

$$R_{Cu} = \frac{\rho_{Cu}L}{A_{Cu}} = R_{Al} = \frac{\rho_{Al}L}{A_{Al}} \Rightarrow A_{Al} = \left(\frac{\rho_{Al}}{\rho_{Cu}}\right)A_{Cu} \Rightarrow \pi r_{Al}^2 = \left(\frac{\rho_{Al}}{\rho_{Cu}}\right)\pi r_{Cu}^2$$

$$\Rightarrow r_{Al} = \sqrt{\frac{\rho_{Al}}{\rho_{Cu}}} r_{Cu} = \sqrt{\frac{2.8 \times 10^{-8} \ \Omega \ m}{1.7 \times 10^{-8} \ \Omega \ m}}(0.25 \ mm) = 0.32 \ mm$$

We need a 0.64-mm-diameter aluminum wire.

30.41. Solve: (a) The current associated with the moving film is the rate at which the charge on the film moves past a certain point. The tangential speed of the film is

$$v = \omega r = (90 \ rpm)(4.0 \ cm) = 90\frac{rev}{min} \times \frac{1 \ min}{60 \ s} \times \frac{2\pi \ rad}{1 \ rev} \times 1.0 \ cm = 9.425 \ cm/s$$

In 1.0 s the film moves a distance of 9.425 cm. This means the area of the film that moves to the right in 1.0 s is $(9.425 \ cm)(4.0 \ cm) = 37.7 \ cm^2$. The amount of charge that passes to the right in 1.0 s is

$$Q = (37.7 \ cm^2)(-2.0 \times 10^{-9} \ C/cm^2) = -75.4 \times 10^{-9} \ C$$

Since $I = Q/\Delta t$, we have

$$I = \frac{\left|-(75.4 \times 10^{-9} \ C)\right|}{1 \ s} = 75.4 \ nA$$

The current is 75 nA.
(b) Having found the current in part (a), we can once again use $I = Q/\Delta t$ to obtain Δt:

$$\Delta t = \frac{Q}{I} = \frac{\left|-10 \times 10^{-6} \ C\right|}{75.4 \times 10^{-9} \ A} = 133 \ s \approx 130 \ s$$

30.45. Solve: (a) Current is defined as $I = Q/\Delta t$, so the charge delivered in time Δt is
$$Q = I\Delta t = (150 \ A)(0.80 \ s) = 120 \ C$$

(b) The drift speed is

$$v_d = \frac{J}{ne} = \frac{I/A}{ne} = \frac{I}{\pi r^2 ne} = \frac{150 \ A}{\pi(0.0025 \ m)^2(8.5 \times 10^{28} \ m^{-3})(1.60 \times 10^{-19} \ C)} = 5.617 \times 10^{-4} \ m/s$$

At this speed the electrons drift a distance

$$d = (5.617 \times 10^{-4} \ m/s)(0.80 \ s) = 4.49 \times 10^{-4} \ m = 0.45 \ mm$$

30.47. Model: We assume that Ohm's law applies to the situation.

$$I = \frac{\Delta V}{R}$$

We also use Equation 30.22, which gives R in terms of ρ, L, and A.

$$R = \frac{\rho L}{A}$$

Visualize: We are given that $\Delta V = 9.0 \ V$, $L = 0.050 \ m$, $I = 230 \ \mu A$, and $A = \pi r^2 = \pi(d/2)^2 = \pi(1.5 \ mm/2)^2 = 1.77 \times 10^{-6} \ m^2$.

Solve: Combine the two previous equations.

$$\rho = \frac{RA}{L} = \frac{\Delta V}{I}\frac{A}{L} = \frac{(9.0 \text{ V})(1.77 \times 10^{-6} \text{ m}^2)}{(230 \times 10^{-6} \text{ A})(0.050 \text{ m})} = 1.4 \text{ } \Omega \cdot \text{m}$$

Assess: This resistivity is close to the value for blood ($1.6 \text{ } \Omega \cdot \text{m}$) found in references.

30.51. Visualize:

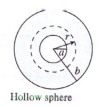

Hollow sphere

Solve: (a) Consider a spherical surface inside the hollow sphere at a radial distance r from the center. The current is flowing outward through this surface, which has surface area $A = 4\pi r^2$. Thus

$$I = JA = \sigma E(4\pi r^2)$$

Thus the electric field strength at radius r is

$$E = \frac{I}{4\pi \sigma r^2}$$

(b) For copper, with $\sigma = 1.0 \times 10^7 \text{ } \Omega^{-1} \text{ m}^{-1}$,

$$E_{\text{inner}} = \frac{25 \text{ A}}{4\pi(6.0 \times 10^7 \text{ } \Omega^{-1} \text{ m}^{-1})} \frac{1}{(0.010 \text{ m})^2} = 3.3 \times 10^{-4} \text{ V/m}$$

$$E_{\text{outer}} = \frac{25 \text{ A}}{4\pi(6.0 \times 10^7 \text{ } \Omega^{-1} \text{ m}^{-1})} \frac{1}{(0.025 \text{ m})^2} = 5.3 \times 10^{-5} \text{ N/C}$$

30.53. Solve: (a) Since $I = \Delta Q/\Delta t$, for infinitesimal changes

$$I = \frac{dQ}{dt} = \frac{d}{dt}(20 \text{ C})(1 - e^{-t/2.0 \text{ s}}) = (20 \text{ C})(-e^{-t/2.0 \text{ s}})(-1/2.0 \text{ s}) = (10 \text{ A})e^{-t/2.0 \text{ s}}$$

(b) The maximum value of the current occurs at $t = 0$ s and is 10 A.
(c) The values of I (A) for selected values of t are

t (s)	0	1	2	4	6	8	10
I (A)	10	6.07	3.68	1.35	0.50	0.18	0.07

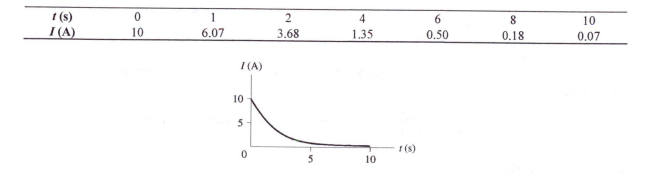

30.55. Model: Because current is conserved, the current flowing in the 2.0-mm-diameter segment of the wire is the same as in the 1.0-mm-diameter segment.
Visualize: We will denote all quantities for the 1.0-mm-diameter wire with the subscript 1, and all quantities for the 2.0-mm-diameter wire with the subscript 2.
Solve: Equation 30.13 is $J = nev_d$. This means the current densities in the two segments are

$$J_1 = nev_{d1} \qquad J_2 = nev_{d2}$$

Dividing these equations, we get $v_{d2} = (J_2/J_1)v_{d1}$. Because current is conserved, $I_1 = I_2 = 2.0$ A. So,

$$\frac{J_2}{J_1} = \frac{I_2/A_2}{I_1/A_1} = \frac{A_1}{A_2} \Rightarrow v_{d2} = \frac{A_1}{A_2}v_{d1} = \left(\frac{D_1}{D_2}\right)^2 v_{d1} = \left(\frac{1.0 \text{ mm}}{2.0 \text{ mm}}\right)^2 (2.0 \times 10^{-4} \text{ m/s}) = 5.0 \times 10^{-5} \text{ m/s}$$

Assess: A drift velocity which is small and only $\left(\frac{1}{4}\right)$ of the drift velocity in the 1.0-mm-diameter wire is reasonable.

30.59. Solve: From Equation 30.17 and Table 30.2, the electric field is

$$E = \frac{J}{\sigma} = \frac{I}{\sigma A} = \frac{5.0 \text{ A}}{(1.0 \times 10^7 \text{ } \Omega^{-1}\text{m}^{-1})\pi(1.0 \times 10^{-3} \text{ m})^2} = 0.16 \text{ V/m}$$

30.61. Model: Assume the resistors are ohmic.
Solve: The potential difference across the resistor on the left is ε, so the current in it is $I = \varepsilon/R$. The potential difference across the resistor on the right is 2ε. For the current to be the same in that resistor, the resistance must be twice as much, so that $I = 2\varepsilon/(2R)$. Therefore, the resistance of the resistor on the right is $2R$.
Assess: The two resistors are neither in parallel nor in series.

30.63. Model: The charged metal plates form a parallel-plate capacitor.
Solve: (a) When the charged plates are connected, the maximum current in the wire can be obtained from $I_{max} = \Delta V_{max}/R$ where ΔV_{max} is the potential difference across the capacitor just before the wire is connected. ΔV_{max} for the capacitor is $\Delta V_{max} = Q/C$, where C is the capacitance for a parallel-plate capacitor. Noting that $C = \varepsilon_0 A/d$ and $R = \rho L/\pi r^2$, the maximum current is

$$I_{max} = \frac{\Delta V_{max}}{R} = \frac{Q}{CR} = \frac{Q}{(\varepsilon_0 A/d)(\rho L/\pi r^2)} = \frac{Q}{\varepsilon_0 \rho} \frac{\pi r^2}{A}$$

$$= \frac{(12.5 \times 10^{-9} \text{ C})\pi(0.112 \times 10^{-3} \text{ m})^2}{(8.85 \times 10^{-12} \text{ C}^2/\text{N m}^2)(1.7 \times 10^{-8} \text{ } \Omega \text{ m})\pi(5.0 \times 10^{-2} \text{ m})^2} = 4.2 \times 10^5 \text{ A}$$

In the above calculations, note that the length of the wire L is the same as the separation of the two plates d. Also, the resisitivity of copper is taken from Table 30.2.
(b) The current I will decrease in time. As charge leaves one plate and moves to the other plate, the voltage across the capacitor and hence the current goes down.
(c) The energy dissipated in the wire is the energy present in the capacitor before it was connected with the wire. This energy is

$$U_{dissipated} = \frac{1}{2}\frac{Q^2}{C} = \frac{1}{2}Q\frac{Q}{C} = \frac{1}{2}Q\Delta V_{max} = \frac{1}{2}(12.5 \times 10^{-9} \text{ C})(1800 \text{ V}) = 1.1 \times 10^{-5} \text{ J}$$

30.67. Solve: (a) The charge delivered is

$$(50 \times 10^3 \text{ A})(50 \times 10^{-6} \text{ s}) = 2.5 \text{ C}.$$

(b) The current in the lightning rod and the potential drop across it are related by Equation 30.21. Using ρ for iron from Table 30.2,

$$I = \frac{A}{\rho L}\Delta V \Rightarrow A = \frac{\rho L I}{\Delta V} = \frac{(9.7\times10^{-8}\ \Omega m)(5.0\ m)(50\times10^{3}\ A)}{100\ V} = 2.43\times10^{-4}\ m^{2}$$

This is the area required for a maximum voltage drop of 100 V. The corresponding diameter of the lightning rod is

$$d = 2r = 2\sqrt{\frac{A}{\pi}} = 2\sqrt{\frac{2.43\times10^{-4}\ m^{2}}{\pi}} = 1.8\times10^{-2}\ m = 1.8\ cm$$

FUNDAMENTALS OF CIRCUITS

Exercises and Problems

Section 31.1 Circuit Elements and Diagrams

Section 31.2 Kirchhoff's Laws and the Basic Circuit

31.3. Model: Assume that the connecting wires are ideal.
Visualize: Please refer to Figure EX31.3.
Solve: The current in the 2 Ω resistor is $I_1 = 4$ V/2 Ω $= 2$ A to the right. The current in the 5 Ω resistor is $I_2 = (15$ V)/ $(5 \, \Omega) = 3$ A downward. Let I be the current flowing out of the junction. Kirchhoff's junction law then gives

$$I = I_1 - I_2 = 2 \text{ A} - 3 \text{ A} = -1 \text{ A}$$

Thus, 1 A of current flows into the junction (i.e., to the left).

31.5. Model: The batteries and the connecting wires are ideal.
Visualize: Please refer to Figure EX31.5.
Solve: **(a)** Choose the current I to be in the clockwise direction. If I ends up being a positive number, then the current really does flow in this direction. If I is negative, the current really flows counterclockwise. There are no junctions, so I is the same for all elements in the circuit. With the 9 V battery labeled 1 and the 6 V battery labeled 2, Kirchhoff's loop law gives

$$\Sigma \Delta V_i = \Delta V_{\text{bat 1}} + \Delta V_R + \Delta V_{\text{bat 2}} = +\mathcal{E}_1 - IR - \mathcal{E}_2 = 0$$
$$I = \frac{\mathcal{E}_1 - \mathcal{E}_2}{R} = \frac{9 \text{ V} - 18 \text{ V}}{10 \, \Omega} = -0.9 \text{ A}$$

Note the signs: Potential is gained in battery 1, but potential is lost both in the resistor and in battery 2. Because I is negative, we can say that $I = 0.9$ A and flows from right to left through the resistor.
(b) In the graph below, we start at the lower-left corner of the circuit and travel clockwise around the circuit (i.e., against the current). We start by losing 9 V going through battery 1, then loss $\Delta V_R = -IR = 9$ V going through the resistor. We then gain 18 V going through battery 2. The final potential is the same as the initial potential, as required.

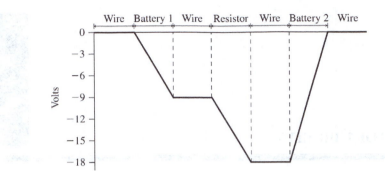

Section 31.3 Energy and Power

31.9. Model: Assume ideal connecting wires and that both bulbs have the same rated voltage (normally 120 V in the United States, although this value does not matter for this problem).
Visualize: Please refer to Figure EX31.9.
Solve: If the bulbs are operated at their rated voltage V_{Rating}, their resistance is

$$60 \text{ W} = (\Delta V_{\text{Rating}})^2 / R_{60} \quad \Rightarrow \quad R_{60} = \frac{(\Delta V_{\text{Rating}})^2}{60 \text{ W}}$$

$$100 \text{ W} = (\Delta V_{\text{Rating}})^2 / R_{100} \quad \Rightarrow \quad R_{100} = \frac{(\Delta V_{\text{Rating}})^2}{100 \text{ W}}$$

When operated with the battery of unknown voltage, the power output of the bulbs will be different. We shall call these powers P_{60} and P_{100}. We know that the same current must run through both bulbs, so we can express their power output in terms of their known resistance and the current I:

$$\left.\begin{array}{c} P_{60} = I^2 R_{60} \\ P_{100} = I^2 R_{100} \end{array}\right\} \quad \Rightarrow \quad \frac{P_{60}}{P_{100}} = \frac{I^2 R_{60}}{I^2 R_{100}} = \frac{(\Delta V_{\text{Rating}})^2/(60 \text{ W})}{(\Delta V_{\text{Rating}})^2/(100 \text{ W})} = \frac{100 \text{ W}}{60 \text{ W}} > 1$$

Therefore, the 60 W bulb emits more power (or is brighter) than the 100 W bulb, so the response is A.

31.11. Solve: (a) The average power consumed by a typical American family is

$$P_{\text{avg}} = 1000 \frac{\text{kWh}}{\text{month}} = 1000 \frac{\text{kWh}}{30 \times 24 \text{ h}} = \frac{1000}{720} \text{ kW} = 1.389 \text{ kW}$$

Because $P = (\Delta V)I$ with ΔV being the voltage of the power line to the house,

$$I_{\text{avg}} = \frac{P_{\text{avg}}}{\Delta V} = \frac{1389 \text{ W}}{120 \text{ V}} = 11.6 \text{ A}$$

(b) Because $P = (\Delta V)^2 / R$,

$$R_{\text{avg}} = \frac{(\Delta V)^2}{P_{\text{avg}}} = \frac{(120 \text{ V})^2}{1389 \text{ W}} = 10.4 \text{ } \Omega$$

Section 31.4 Series Resistors

Section 31.5 Real Batteries

31.13. Visualize: Please refer to Figure EX31.13.
Solve: The three resistors are in series. The total resistance is 200 Ω, so the unknown resistance R is

$$R_{\text{eq}} = R + 50 \text{ } \Omega + R = 2R + 50 \text{ } \Omega \quad \Rightarrow \quad R = \frac{R_{\text{eq}} - 50 \text{ } \Omega}{2} = \frac{200 \text{ } \Omega - 50 \text{ } \Omega}{2} = 75 \text{ } \Omega$$

31.17. Model: Assume ideal connecting wires but not an ideal battery.

Visualize: The circuit for an ideal battery is the same as the circuit in Figure EX31.17, except that the 1 Ω resistor is not present.

Solve: In the case of an ideal battery, we have a battery with $\mathcal{E} = 15$ V connected to two series resistors of 10 Ω and 20 Ω resistance. Because the equivalent resistance is $R_{eq} = 10\ \Omega + 20\ \Omega = 30\ \Omega$ and the potential difference across R_{eq} is 15 V, the current in the circuit is

$$I = \frac{\Delta V}{R_{eq}} = \frac{\mathcal{E}}{R_{eq}} = \frac{15\text{ V}}{30\ \Omega} = 0.50\text{ A}$$

The potential difference across the 20 Ω resistor is

$$\Delta V_{20} = IR = (0.50\text{ A})(20\ \Omega) = 10\text{ V}$$

In the case of a real battery, we have a battery with $\mathcal{E} = 15$ V connected to three series resistors: 10 Ω, 20 Ω, and an internal resistance of 1 Ω. Now the equivalent resistance is

$$R'_{eq} = 10\ \Omega + 20\ \Omega + 1\ \Omega = 31\ \Omega$$

The potential difference across R_{eq} is the same as before $\mathcal{E} = 15$ V. Thus,

$$I' = \frac{\Delta V'}{R'_{eq}} = \frac{\mathcal{E}}{R'_{eq}} = \frac{15\text{ V}}{31\ \Omega} = 0.4839\text{ A}$$

Therefore, the potential difference across the 20 Ω resistor is

$$\Delta V'_{20} = I'R = (0.4839\text{ A})(20\ \Omega) = 9.68\text{ V}$$

That is, the potential difference across the 20 Ω resistor is reduced from 10 V to 9.68 V due to the internal resistance of 1 Ω of the battery. The percentage change in the potential difference is

$$\left(\frac{10.0\text{ V} - 9.68\text{ V}}{10.0\text{ V}} \right) \times 100\% = 3.2\%$$

Section 31.6 Parallel Resistors

31.19. Visualize: The three resistors in Figure EX31.19 are equivalent to a resistor of resistance $R_{eq} = 75\ \Omega$.

Solve: Because the three resistors are in parallel,

$$\frac{1}{R_{eq}} = \frac{1}{R} + \frac{1}{200\ \Omega} + \frac{1}{R} = \frac{2}{R} + \frac{1}{200\ \Omega} = \frac{400\ \Omega + R}{(200\ \Omega)R} \quad \Rightarrow \quad R_{eq} = 75\ \Omega = \frac{(200\ \Omega)R}{(400\ \Omega + R)} = \frac{200\ \Omega}{1 + \left(\dfrac{400\ \Omega}{R} \right)}$$

$$R = \frac{400\ \Omega}{\dfrac{200\ \Omega}{75\ \Omega} - 1} = 240\ \Omega$$

31.21. Model: The connecting wires are ideal with zero resistance.

Solve:

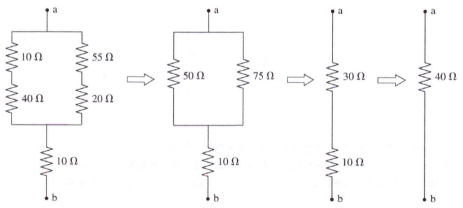

For the first step, the 10 Ω and 40 Ω resistors are in series and the equivalent resistance is 50 Ω. Likewise, the 55 Ω and 20 Ω resistors give an equivalent resistance of 75 Ω. For the second step, the 75 Ω and 50 Ω resistors are in parallel and the equivalent resistance is

$$\left[\frac{1}{50\ \Omega} + \frac{1}{75\ \Omega} \right]^{-1} = 30\ \Omega$$

For the third step, the 30 Ω and 10 Ω resistors are in series and the equivalent resistance is 40 Ω.

Section 31.8 Getting Grounded

Section 31.9 *RC* Circuits

31.27. Solve: Noting that the unit of resistance is the ohm (V/A) and the unit of capacitance is the farad (C/V), the unit of *RC* is

$$RC = \frac{V}{A} \times \frac{C}{V} = \frac{C}{A} = \frac{C}{C/s} = s$$

31.29. Model: Assume ideal wires as the capacitors discharge through the two 1 kΩ resistors.
Visualize: The circuit in Figure EX31.29 has an equivalent circuit with resistance R_{eq} and capacitance C_{eq}.
Solve: The equivalent capacitance is

$$\frac{1}{C_{eq}} = \frac{1}{2\ \mu F} + \frac{1}{2\ \mu F} \quad \Rightarrow \quad C_{eq} = 1\ \mu F$$

and the equivalent resistance is $R_{eq} = 1\ k\Omega + 1\ k\Omega = 2\ k\Omega$. Thus, the time constant for the discharge of the capacitors is

$$\tau = R_{eq} C_{eq} = (2\ k\Omega)(1\ \mu F) = 2 \times 10^{-3}\ s = 2\ ms$$

Assess: The capacitors will be almost entirely discharged $5\tau = 5 \times 2\ ms = 10\ ms$ after the switch is closed.

31.33. Model: A capacitor discharges through a resistor. Assume ideal wires.
Solve: The discharge current or the resistor current follows Equation 31.35: $I = I_0 e^{-t/RC}$. We wish to find the capacitance *C* so that the resistor current will decrease to 25% of its initial value in 2.5 ms. That is,

$$0.25 I_0 = I_0 e^{-(2.5\ ms)/(100\ \Omega)C} \quad \Rightarrow \quad \ln(0.25) = -\frac{2.5 \times 10^{-3}\ s}{(100\ \Omega)C} \quad \Rightarrow \quad C = 18\ \mu F$$

31.35. Model: Assume ideal wires.
Visualize: The following equivalent circuits will be useful, where we have labeled three points in the circuits a, b, and c. The unlabeled resistors all have resistance *R*.

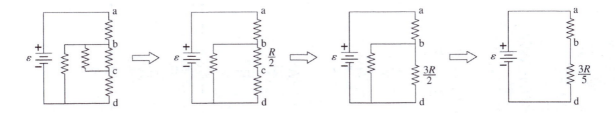

Solve: Because the bulbs are identical, they all have the same resistance, which we shall call *R*. The bulb brightness is proportional to the power $P = V^2/R$, so we shall calculate the voltage difference across each bulb. First, we find the potential at points a, b, and c with respect to point d (i.e., the negative terminal of the battery), which we shall assign as the zero of the potential. The potential at point a is ε, so $V_a = \varepsilon$. Considering the last equivalent circuit above, we see that the current flowing around the circuit is

$$I = \frac{V_a}{R + 3R/5} = \frac{5\varepsilon}{8R}$$

Therefore, the potential at point b is

$$V_b = V_a - IR = \varepsilon - \frac{5\varepsilon}{8} = \frac{3\varepsilon}{8}$$

Considering the second-to-last circuit, we see that the current flowing from point b to point d must be

$$I_R = \frac{V_b}{3R/2} = \frac{2}{3R} \frac{3\varepsilon}{8} = \frac{\varepsilon}{4R}$$

so, looking at the second circuit, we can find the potential at point c:

$$V_c = V_b - I_R\left(\frac{R}{2}\right) = \frac{3\varepsilon}{8} - \left(\frac{\varepsilon}{4R}\right)\left(\frac{R}{2}\right) = \frac{\varepsilon}{4}$$

In the following table, we put the bulbs and the potential difference across them:

Bulb	Potential Difference	Power
P	$\Delta V = V_b - V_a = -\dfrac{5\varepsilon}{8}$	$\dfrac{25}{64}\dfrac{\varepsilon^2}{R}$
Q	$\Delta V = V_d - V_b = -\dfrac{3\varepsilon}{8}$	$\dfrac{9}{64}\dfrac{\varepsilon^2}{R}$
R	$\Delta V = V_c - V_b = -\dfrac{\varepsilon}{8}$	$\dfrac{1}{64}\dfrac{\varepsilon^2}{R}$
S	$\Delta V = V_c - V_b = -\dfrac{\varepsilon}{8}$	$\dfrac{1}{64}\dfrac{\varepsilon^2}{R}$
T	$\Delta V = V_d - V_c = -\dfrac{\varepsilon}{4}$	$\dfrac{4}{64}\dfrac{\varepsilon^2}{R}$

Thus, ordering the bulbs from brightest to dimmest gives

$$P > Q > T > R = S$$

so the response is D.

31.37. Model: Assume ideal connecting wires and an ideal power supply.
Visualize:

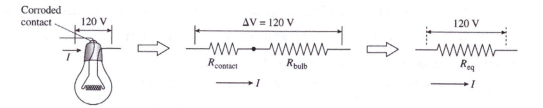

Solve: We have two resistors in series such that $R_{eq} = R_{bulb} + R_{contacts}$. R_{bulb} can be found from the fact that we have a 100 W (120 V) bulb:

$$R_{bulb} = \frac{\Delta V^2}{P} = \frac{(120 \text{ V})^2}{100 \text{ W}} = 144 \text{ } \Omega$$

We have a total resistance of $R_{eq} = 144 \text{ } \Omega + 5.0 \text{ } \Omega = 149 \text{ } \Omega$. The current flowing through R_{eq} is

$$I = \frac{\Delta V}{R_{eq}} = \frac{120 \text{ V}}{149 \text{ } \Omega} = 0.8054 \text{ A}$$

Because R_{eq} is a series combination of R_{bulb} and $R_{contacts}$, this current flows through both the bulb and the contact. Thus,

$$P_{bulb} = I^2 R_{bulb} = (0.8054 \text{ A})^2 (144 \text{ } \Omega) = 93 \text{ W}$$

Assess: The corroded contact changes the circuit's total resistance and reduces the current below that at which the bulb was rated. So, it makes sense for it to operate at less than full power.

31.39. Model: The wires and battery are ideal.
Visualize:

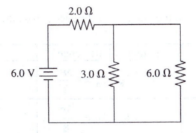

Solve: We can find the equivalent resistance necessary for the battery to deliver 9 W of power:

$$P = \frac{(\Delta V)^2}{R} \quad \Rightarrow \quad R = \frac{(\Delta V)^2}{P} = \frac{(6.0 \text{ V})^2}{9.0 \text{ W}} = 4.0 \text{ } \Omega$$

The combination of the 2.0 Ω, 3.0 Ω, and 6.0 Ω resistors that make 4.0 Ω is shown in the figure above. The 3.0 Ω and 6.0 Ω parallel combination has an equivalent resistance of 2.0 Ω, which when added to the 2.0 Ω resistor in series totals 4.0 Ω equivalent resistance.

31.45. Model: Assume that the connecting wires are ideal, but the battery is not. The battery has internal resistance. Also assume that the ammeter does not have any resistance.
Visualize: Please refer to Figure P31.45.
Solve: When the switch is open,

$$\varepsilon - Ir - I(5.0 \text{ } \Omega) = 0 \quad \Rightarrow \quad V = (1.636 \text{ A})(r + 5.0 \text{ } \Omega)$$

where we applied Kirchhoff's loop law starting from the lower-left corner. When the switch is closed, the current I comes out of the battery and splits at the junction. The current $I' = 1.565$ A flows through the 5.0 Ω resistor and the rest $(I - I')$ flows through the 10.0 Ω resistor. Because the potential differences across the two resistors are equal,

$$I'(5.0 \text{ } \Omega) = (I - I')(10.0 \text{ } \Omega) \quad \Rightarrow \quad (1.565 \text{ A})(5.0 \text{ } \Omega) = (I - 1.565 \text{ A})(10.0 \text{ } \Omega) \quad \Rightarrow \quad I = 2.348 \text{ A}$$

Applying Kirchhoff's loop law to the left loop of the closed circuit gives

$$\varepsilon - Ir - I'(5.0 \text{ } \Omega) = 0 \text{ V} \quad \Rightarrow \quad \varepsilon = (2.348 \text{ A})r + (1.565 \text{ A})(5.0 \text{ } \Omega) = (2.348 \text{ A})r + 7.825 \text{ V}$$

Combining this equation for ε with the equation obtained from the circuit when the switch was open gives

$$(2.348 \text{ A})r + 7.825 \text{ V} = (1.636 \text{ A})r + 8.18 \text{ V} \quad \Rightarrow \quad (0.712 \text{ A})r = 0.355 \text{ V} \quad \Rightarrow \quad r = 0.50 \text{ } \Omega$$

We also have $\varepsilon = (1.636 \text{ A})(0.50 \text{ } \Omega + 5.0 \text{ } \Omega) = 9.0 \text{ V}$.

31.47. Model: The connecting wires are ideal.
Visualize:

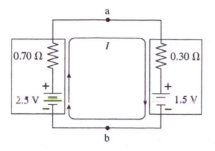

Solve: Let the current in the circuit be I. The terminal voltage of the 2.5 V battery is $V_a - V_b$. This is also the terminal voltage of the 1.5 V battery. $V_a - V_b$ can be obtained by noting that

$$V_b + 2.5\ \text{V} - I(0.70\ \Omega) = V_a \quad \Rightarrow \quad V_a - V_b = 2.5\ \text{V} - I(0.70\ \Omega)$$

To determine I, we apply Kirchhoff's loop law starting from the lower-left corner:

$$+2.5\ \text{V} - I(0.70\ \Omega) - I(0.30\ \Omega) - 1.5\ \text{V} = 0\ \text{V} \quad \Rightarrow \quad I = 1.0\ \text{A}$$

Thus, $V_a - V_b = 2.5\ \text{V} - (1.0\ \text{A})(0.70\ \Omega) = 1.8\ \text{V}$. That is, the terminal voltage of the 1.5 V and the 2.5 V batteries is 1.8 V.

31.49. Model: The batteries are ideal, the connecting wires are ideal, and the ammeter has a negligibly small resistance.
Visualize: Please refer to Figure P31.49.
Solve: Kirchhoff's junction law tells us that the current flowing through the 2.0 Ω resistance in the middle branch is $I_1 + I_2 = 3.0\ \text{A}$. We can therefore determine I_1 by applying Kirchhoff's loop law to the left loop. Starting clockwise from the lower-left corner,

$$+9.0\ \text{V} - I_1(3.0\ \Omega) - (3.0\ \text{A})(2.0\ \Omega) = 0\ \text{V} \quad \Rightarrow \quad I_1 = 1.0\ \text{A}$$
$$I_2 = (3.0\ \text{A} - I_1) = (3.0\ \text{A} - 1.0\ \text{A}) = 2.0\ \text{A}$$

Finally, to determine the emf ε, we apply Kirchhoff's loop law to the right loop and start counterclockwise from the lower-right corner of the loop:

$$+\varepsilon - I_2(4.5\ \Omega) - (3.0\ \text{A})(2.0\text{W}) = 0\ \text{V} \quad \Rightarrow \quad \varepsilon - (2.0\ \text{A})(4.5\ \Omega) - 6.0\ \text{V} = 0\ \text{V} \quad \Rightarrow \quad \varepsilon = 15\ \text{V}$$

31.51. Solve: Let the units guide you. For the incandescent bulb, the life-cycle cost p_{bulb} is

$$p_{\text{bulb}} = \$0.50 + (0.060\ \text{kW})(0.10\ \$/\text{kW h})(1{,}000\ \text{h}) = \$6.50$$

This will give 1,000 hours, so the cost for 10,000 hours is $65.00. For the fluorescent tube, the cost for 10,000 hours is

$$p_{\text{tube}} = \$5 + (0.015\ \text{kW})(0.10\ \$/\text{kW h})(10{,}000\ \text{h}) = \$20$$

Assess: The lifetime cost of the fluorescent bulb is one-third that of the incandescent bulb.

31.55. Model: The voltage source and the connecting wires are ideal.
Visualize: Please refer to Figure P31.55.
Solve: Let us first apply Kirchhoff's loop law starting clockwise from the lower-left corner:

$$+V_{\text{in}} - IR - I(100\ \Omega) = 0\ \text{V} \quad \Rightarrow \quad I = \frac{V_{\text{in}}}{R + 100\ \Omega}$$

The output voltage is

$$V_{\text{out}} = (100\ \Omega)I = (100\ \Omega)\left(\frac{V_{\text{in}}}{R + 100\ \Omega}\right) \quad \Rightarrow \quad \frac{V_{\text{out}}}{V_{\text{in}}} = \frac{100\ \Omega}{R + 100\ \Omega}$$

For $V_{out} = V_{in}/10$, the above equation can be simplified to obtain R:

$$\frac{V_{in}/10}{V_{in}} = \frac{100\ \Omega}{R + 100\ \Omega} \quad \Rightarrow \quad R + 100\ \Omega = 1000\ \Omega \quad \Rightarrow \quad R = 900\ \Omega$$

31.59. Model: The battery and the connecting wires are ideal.
Visualize:

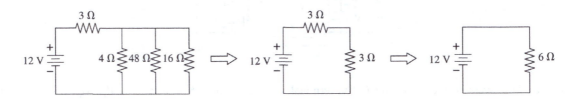

The figure shows how to simplify the circuit in Figure P31.59 using the laws of series and parallel resistances. Having reduced the circuit to a single equivalent resistance, we will reverse the procedure and "build up" the circuit using the loop law and the junction law to find the current and potential difference across each resistor.
Solve: From the last circuit in the diagram,

$$I = \frac{\varepsilon}{6\ \Omega} = \frac{12\ V}{6\ \Omega} = 2\ A$$

Thus, the current through the battery is 2 A. As we rebuild the circuit, we note that series resistors *must* have the same current I and that parallel resistors *must* have the same potential difference ΔV.

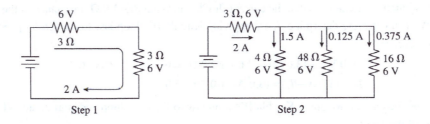

In Step 1, the 6 Ω resistor is returned to a 3 Ω and 3 Ω resistor in series. Both resistors must have the same 2 A current as the 6 Ω resistance. We then use Ohm's law to find

$$\Delta V_3 = (2\ A)(3\ \Omega) = 6\ V$$

As a check, $6\ V + 6\ V = 12\ V$, which was ΔV of the 6 Ω resistor. In Step 2, one of the two 3 Ω resistances is returned to the 4 Ω, 48 Ω, and 16 Ω resistors in parallel. The three resistors must have the same $\Delta V = 6\ V$. From Ohm's law,

$$I_4 = \frac{6\ V}{4\ \Omega} = 1.5\ A \qquad I_{48} = \frac{6\ V}{48\ \Omega} = \frac{1}{8}\ A \qquad I_{16} = \frac{6\ V}{16\ \Omega} = \frac{3}{8}\ A$$

Resistor	Potential difference (V)	Current (A)
3 Ω	6	2
4 Ω	6	1.5
48 Ω	6	1/8
16 Ω	6	3/8

Assess: Note that the currents flowing through the three parallel resistors sum to the 2 A flowing through the battery, as required.

31.61. Model: The batteries and the connecting wires are ideal.
Visualize:

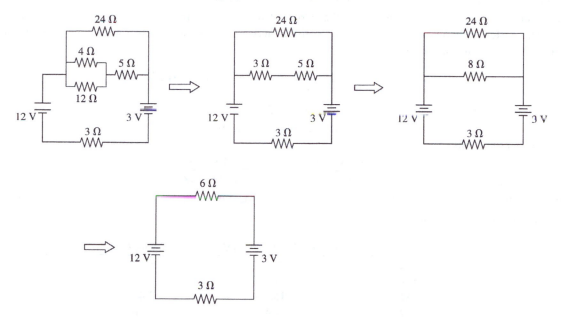

The figure shows how to simplify the circuit in Figure P31.61 using the laws of series and parallel resistances. Having reduced the circuit to a single equivalent resistance, we will reverse the procedure and "build up" the circuit using the loop law and the junction law to find the current and potential difference of each resistor.
Solve: From the last circuit in the figure and from Kirchhoff's loop law,

$$I = \frac{12 \text{ V} - 3 \text{ V}}{6 \text{ }\Omega + 3 \text{ }\Omega} = 1 \text{ A}$$

Thus, the current through the batteries is 1 A. As we rebuild the circuit, we note that series resistors *must* have the same current I and that parallel resistors *must* have the same potential difference.

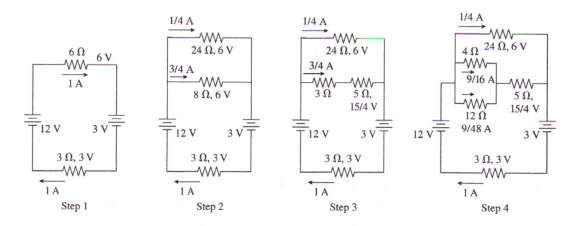

In Step 1 of the above figure, both resistors must have the same 1 A current. We use Ohm's law to find

$$\Delta V_3 = (1 \text{ A})(3 \text{ }\Omega) = 3 \text{ V} \qquad \Delta V_{6\text{eq}} = 6 \text{ V}$$

As a check we sum the voltages around the circuit starting at the lower-left corner: $12\text{V} - 6 \text{ V} - 3 \text{ V} - 3 \text{ V} = 0 \text{ V}$, as required. In Step 2, the $6 \text{ }\Omega$ equivalent resistor is returned to the $24 \text{ }\Omega$ and $8 \text{ }\Omega$ resistors in parallel. The two resistors must have the same potential difference $\Delta V = 6 \text{ V}$. From Ohm's law,

$$I_{8eq} = \frac{6 \text{ V}}{8 \text{ }\Omega} = \frac{3}{4} \text{ A} \qquad I_{24} = \frac{6 \text{ V}}{24 \text{ }\Omega} = \frac{1}{4} \text{ A}$$

As a check, $3/4 \text{ A} + 1/4 \text{ A} = 1 \text{ A}$ which was the current I of the $6 \text{ }\Omega$ equivalaent resistor. In Step 3, the $8 \text{ }\Omega$ equivalent resistor is returned to the $3 \text{ }\Omega$ and $5 \text{ }\Omega$ resistors in series, so the two resistors must have the same current of $3/4 \text{ A}$. We use Ohm's law to find

$$\Delta V_{3eq} = \left(\frac{3}{4} \text{ A}\right)(3 \text{ }\Omega) = \frac{9}{4} \text{ V} \qquad \Delta V_5 = \left(\frac{3}{4} \text{ A}\right)(5 \text{ }\Omega) = \frac{15}{4} \text{ V}$$

As a check, $9/4 \text{ V} + 15/4 \text{ V} = 24/4 \text{ V} = 6 \text{ V}$, which was ΔV of the $8 \text{ }\Omega$ equivalent resistor. In Step 4, the $3 \text{ }\Omega$ equivalent resistor is returned to $4 \text{ }\Omega$ and $12 \text{ }\Omega$ resistors in parallel, so the two must have the same potential difference $\Delta V = 9/4 \text{ V}$. From Ohm's law,

$$I_4 = \frac{9/4 \text{ V}}{4 \text{ }\Omega} = \frac{9}{16} \text{ A} \qquad I_{12} = \frac{9/4 \text{ V}}{12 \text{ }\Omega} = \frac{9}{48} \text{ A}$$

As a check, $9/16 \text{ A} + 9/48 \text{ A} = 3/4 \text{ A}$, which was the same as the current through the $3 \text{ }\Omega$ equivalent resistor. The results are summarized in the table below.

Resistor	Potential difference (V)	Current (A)
24 Ω	6	1/4
3 Ω	3	1
5 Ω	15/4	3/4
4 Ω	9/4	9/16
12 Ω	9/4	9/48

31.63. Model: The wires and batteries are ideal.
Visualize:

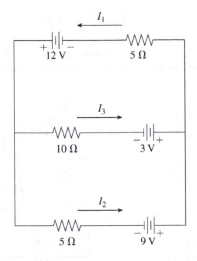

Solve: Assign currents I_1, I_2, and I_3 as shown in the figure above. If I_3 turns out to be negative, we'll know it really flows right to left. Apply Kirchhoff's loop rule counterclockwise to the top loop from the top-right corner:

$$-I_1(5 \text{ }\Omega) + 12 \text{ V} - I_3(10 \text{ }\Omega) + 3 \text{ V} = 0$$

Apply the loop rule counterclockwise to the bottom loop starting at the lower-left corner:

$$-I_2(5 \text{ }\Omega) + 9 \text{ V} - 3 \text{ V} + I_3(10 \text{ }\Omega) = 0$$

Note that, because we went against the current direction through the $10 \text{ }\Omega$ resistor, the potential increased across this resistor. Apply the junction rule to the right middle junction:

$$I_1 = I_2 + I_3.$$

These three equations can be solved for the current I_3:

$$(-I_1 + I_2)(5\,\Omega) + 9\,\text{V} - 2I_3(10\,\Omega) = 0 \quad \Rightarrow \quad -I_3(5\,\Omega) + 9\,\text{V} - 2I_3(10\,\Omega) = 0 \quad \Rightarrow \quad I_3 = \frac{9}{25}\,\text{A}$$

The result is $I_3 = 9/25\,\text{A} = 0.12\,\text{A}$ flowing from left to right (as shown in the figure above).

31.67. Model: The wires and batteries are ideal.
Visualize:

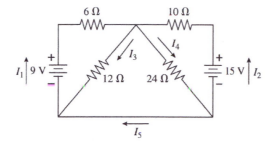

Solve: The circuit is redrawn above for clarity and the currents are shown. We must find I_5.
Repeatedly apply Kirchhoff's rules to the loops. The loop rule applied clockwise about the three triangles yields

$$\text{Left: } 9\,\text{V} - I_1(6\,\Omega) - I_3(12\,\Omega) = 0 \quad \Rightarrow \quad I_1 = 1.5\,\text{A} - 2I_3$$
$$\text{Center: } -I_4(24\,\Omega) + I_3(12\,\Omega) = 0 \quad \Rightarrow \quad I_4 = I_3/2$$
$$\text{Right: } 15\,\text{V} - I_2(10\,\Omega) - I_4(24\,\Omega) = 0 \quad \Rightarrow \quad I_2 = 1.5\,\text{A} - 2.4I_4$$

The junction rule applied at the bottom corners gives equations into which the results above may be substituted:

$$I_1 = I_3 + I_5 \quad \Rightarrow \quad 1.5\,\text{A} - 2I_3 = I_3 + I_5 \quad \Rightarrow \quad I_5 = 1.5\,\text{A} - 3I_3$$
$$I_4 = I_2 + I_5 \quad \Rightarrow \quad I_4 = 1.5\,\text{A} - 2.4I_4 + I_5 \quad \Rightarrow \quad I_5 = 3.4I_4 - 1.5\,\text{A}$$

Using $I_4 = I_3/2$ and solving for I_3 gives

$$1.5\,\text{A} - 3I_3 = 3.4(I_3/2) - 1.5\,\text{A} \quad \Rightarrow \quad I_3 = \frac{30}{47}\,\text{A}$$
$$I_5 = 1.5\,\text{A} - 3\left(\frac{30}{47}\,\text{A}\right) = -\frac{201}{94}\,\text{A} = -0.41\,\text{A}$$

Since the result is negative, 0.40 A flows from left to right through the bottom wire.

31.71. Model: The capacitor discharges through the resistor, and the wires are ideal.
Solve: In an RC circuit, the charge at a given time is related to the original charge as $Q = Q_0 e^{-t/\tau}$. For a capacitor $Q = C\Delta V$, so $\Delta V = \Delta V_0 e^{-t/\tau}$. From the Figure P31.71, we note that $\Delta V_0 = 30\,\text{V}$ and $\Delta V = 10\,\text{V}$ at $t = 4\,\text{ms}$. So,

$$10\,\text{V} = (30\,\text{V})e^{-4\,\text{ms}/R(50\times 10^{-6}\,\text{F})} \quad \Rightarrow \quad \ln\left(\frac{10\,\text{V}}{30\,\text{V}}\right) = -\frac{4\times 10^{-3}\,\text{s}}{R(50\times 10^{-6}\,\text{F})} \quad \Rightarrow \quad R = -\frac{4\times 10^{-3}\,\text{s}}{(50\times 10^{-6}\,\text{F})\ln\left(\frac{1}{3}\right)} = 73\,\Omega$$

THE MAGNETIC FIELD

Exercises and Problems

Section 32.3 The Source of the Magnetic Field: Moving Charges

32.3. Model: The magnetic field is that of a moving charged particle.
Visualize:

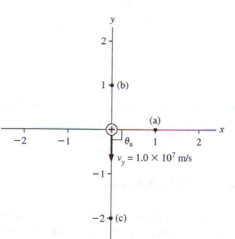

The first point is on the x-axis, with $\theta_a = 0°$. The second point is on the y-axis, with $\theta_b = 90°$, and the third point is on the −y-axis with $\theta_c = 90°$ (in the opposite senese). Use the Bio-Savart law.

$$B = \frac{\mu_0}{4\pi}\frac{qv\sin\theta}{r^2}$$

Solve: (a) $B_a = 0$ T because $\sin\theta_a = \sin 0° = 0$.
(b) Using the Biot-Savart law, the magnetic field strength is

$$B_b = \frac{\mu_0}{4\pi}\frac{qv\sin\theta}{r^2} = \frac{(10^{-7}\ \text{T m/A})(1.60\times10^{-19}\ \text{C})(1.0\times10^{+7}\ \text{m/s})\sin 90°}{(1.0\times10^{-2}\ \text{m})^2} = 1.60\times10^{-15}\ \text{T}$$

To use the right-hand rule for finding the direction of $\vec{B}$, point your thumb in the direction of $\vec{v}$. The magnetic field vector $\vec{B}$ is perpendicular to the plane of $\vec{r}$ and $\vec{v}$ and points in the same direction that your fingers point. In the present case, the fingers point along the $\hat{k}$ direction. Thus, $\vec{B}_b = 1.60\times10^{-15}\hat{k}$ T.

(c) Using the Biot-Savart law, the magnetic field strength is

$$B_c = \frac{\mu_0}{4\pi}\frac{qv\sin\theta}{r^2} = \frac{(10^{-7}\text{ T m/A})(1.60\times10^{-19}\text{ C})(1.0\times10^{+7}\text{ m/s})\sin(90°)}{(2.0\times10^{-2}\text{ m})^2} = 4.0\times10^{-16}\text{ T}$$

To use the right-hand rule for finding the direction of $\vec{B}$, point your thumb in the direction of $\vec{v}$. The magnetic field vector $\vec{B}$ is perpendicular to the plane of $\vec{r}$ and $\vec{v}$ and points in the same direction that your fingers point. In the present case, the fingers point along the $-\hat{k}$ direction. Thus, $\vec{B}_c = -4.0\times10^{-16}\hat{k}$ T.

32.5. Model: The magnetic field is that of a moving charged particle.
Solve: Using the Biot-Savart law,

$$B = \frac{\mu_0}{4\pi}\frac{qv\sin\theta}{r^2} = \frac{(10^{-7}\text{ T m/A})(1.60\times10^{-19}\text{ C})(2.0\times10^7\text{ m/s})\sin45°}{(1.0\times10^{-2}\text{ m})^2 + (1.0\times10^{-2}\text{ m})^2} = 1.13\times10^{-15}\text{ T}$$

The right-hand rule applied to the *proton* points $\vec{B}$ *into* the page. Thus, $\vec{B} = -1.13\times10^{-15}\hat{k}$ T.

32.7. Model: The magnetic field is that of a moving proton.
Visualize:

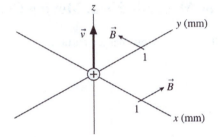

The magnetic field lies in the *xy*-plane.
Solve: Using the right-hand rule, the charge is moving along the +*z*-direction. That is, $\vec{v} = v\hat{k}$. Using the Biot-Savart law,

$$B = \frac{\mu_0}{4\pi}\frac{qv\sin\theta}{r^2} \Rightarrow 1.0\times10^{-13}\text{ T} = \frac{(10^{-7}\text{ T m/A})(1.60\times10^{-19}\text{ C})v\sin90°}{(1.0\times10^{-3}\text{ m})^2} \Rightarrow v = 6.3\times10^6\text{ m/s in the } +z\text{-direction}$$

Section 32.4 The Magnetic Field of a Current

32.9. Model: The magnetic field is that of an electric current in a long, straight wire.
Solve: From Example 32.3, the magnetic field strength of a long, straight wire carrying current I at a distance d from the wire is

$$B = \frac{\mu_0}{2\pi}\frac{I}{d}$$

The distance d at which the magnetic field is equivalent to Earth's magnetic field is calculated as follows:

$$B_{\text{earth surface}} = 5\times10^{-5}\text{ T} = (2\times10^{-7}\text{ T m/A})\frac{10\text{ A}}{d} \Rightarrow d = 4.0\text{ cm}$$

Likewise, the corresponding distances for a refrigerator magnet, a laboratory magnet, and a superconducting magnet are 0.40 mm, 20 μm to 2.0 μm, and 0.20 μm.

Section 32.5 Magnetic Dipoles

32.15. Model: Assume that the 10 cm distance is much larger than the size of the small bar magnet.
Solve: (a) From Equation 32.9, the on-axis field of a magnetic dipole is

$$B = \frac{\mu_0}{4\pi}\frac{2\mu}{z^3} \Rightarrow \mu = \frac{4\pi}{\mu_0}\frac{Bz^3}{2} = \frac{(5.0\times10^{-6}\text{ T})(0.10\text{ m})^3}{2(10^{-7}\text{ T m/A})} = 0.025\text{ A m}^2$$

(b) The on-axis field strength 15 cm from the magnet is

$$B = \frac{\mu_0}{4\pi}\frac{2\mu}{z^3} = (10^{-7}\ \text{T m/A})\frac{2(0.025\ \text{A m}^2)}{(0.15\ \text{m})^3} = 1.48\times10^{-6}\ \text{T} \approx 1.5\,\mu\text{T}$$

32.17. Model: The size of the loop is much smaller than 50 cm, so that the magnetic dipole moment of the loop is $\mu = AI$.
Solve: The magnitude of the on-axis magnetic field of the loop is

$$B = \frac{\mu_0}{4\pi}\frac{2(AI)}{z^3} = 7.5\times10^{-9}\ \text{T} \Rightarrow A = \frac{(0.50\ \text{m})^3(7.5\times10^{-9}\ \text{T})}{2(25\ \text{A})(10^{-7}\ \text{T m/A})} = 1.88\times10^{-4}\ \text{m}^2$$

Let L be the edge-length of the square. Thus

$$L = \sqrt{A} = \sqrt{1.88\times10^{-4}\ \text{m}^2} = 1.37\times10^{-2}\ \text{m} \approx 1.4\ \text{cm}$$

Assess: The 1.4 cm edge-length is much smaller than 50 cm, as we assumed.

Section 32.6 Ampère's Law and Solenoids

32.21. Model: The magnetic field is that of the three currents enclosed by the loop.
Solve: Ampere's law gives the line integral of the magnetic field around the closed path:

$$\oint \vec{B}\cdot d\vec{s} = \mu_0 I_{\text{through}} = 3.77\times10^{-6}\ \text{T m} = \mu_0(I_1 - I_2 + I_3) = (4\pi\times10^{-7}\ \text{T m/A})(2.0\ \text{A} - 6.0\ \text{A} + I_3)$$

$$\Rightarrow (I_3 - 4.0\ \text{A}) = \frac{3.77\times10^{-6}\ \text{T m}}{4\pi\times10^{-7}\ \text{T m/A}} \Rightarrow I_3 = 7.00\ \text{A}$$

Assess: The right-hand rule was used above to assign positive signs to I_1 and I_3 and a negative sign to I_2.

32.23. Model: The magnetic field is that of a current flowing into the plane of the paper. The current-carrying wire is very long.
Solve: Divide the line integral into three parts:

$$\int_i^f \vec{B}\cdot d\vec{s} = \underbrace{\int \vec{B}\cdot d\vec{s}}_{\text{left line}} + \underbrace{\int \vec{B}\cdot d\vec{s}}_{\text{semicircle}} + \underbrace{\int \vec{B}\cdot d\vec{s}}_{\text{right line}}$$

The magnetic field of the current-carrying wire is tangent to clockwise circles around the wire. $\vec{B}$ is everywhere perpendicular to the left line and to the right line, thus the first and third parts of the line integral are zero. Along the semicircle, $\vec{B}$ is tangent to the path *and* has the same magnitude $B = \mu_0 I/2\pi d$ at every point. Thus

$$\int_i^f \vec{B}\cdot d\vec{s} = 0 + BL + 0 = \frac{\mu_0 I}{2\pi d}(\pi d) = \frac{\mu_0 I}{2} = \frac{(4\pi\times10^{-7}\ \text{T m/A})(2.0\ \text{A})}{2} = 1.26\times10^{-6}\ \text{T m}$$

where $L = \pi d$ is the length of the semicircle, which is half the circumference of a circle of radius d.

32.25. Model: Model the solenoid as ideal; the turns of wire are wrapped very closely to each other.
Visualize: The magnetic field in an ideal solenoid does not depend on the cross-sectional area of the solenoid.
Solve: The magnetic field is zero outside an ideal solenoid; inside it is $B_{\text{solenoid}} = \mu_0 nI$ where $n = N/l$ is the number of turns per unit length. The diameter of the wire is the width of one turn which is $l/N = 1/n$.

$$\frac{1}{n} = \frac{\mu_0 I}{B_{\text{solenoid}}} = \frac{(4\pi\times10^{-7}\ \text{T}\cdot\text{m/A})(2.5\ \text{A})}{3.0\ \text{mT}} = 1.0\ \text{mm}$$

Assess: A wire with a width of 1.0 mm seems reasonable. The units check out. We did not need to use the diameter of the solenoid nor its length.

Section 32.7 The Magnetic Force on a Moving Charge

32.27. Model: A magnetic field exerts a magnetic force on a moving charge.
Solve: (a) The force is

$$\vec{F}_{\text{on q}} = q\vec{v} \times \vec{B} = (-1.60 \times 10^{-19} \text{ C})(-1.0 \times 10^{7} \, \hat{k} \text{ m/s}) \times (0.50 \hat{i} \text{ T}) = 8.0 \times 10^{-13} \, \hat{j} \text{ N}$$

(b) The force is

$$\vec{F}_{\text{on q}} = (-1.60 \times 10^{-19} \text{ C})(1.0 \times 10^{7} \text{ m/s})(-\cos 45° \hat{j} + \sin 45° \hat{k}) \times (0.50 \hat{i} \text{ T}) = 5.7 \times 10^{-13} (-\hat{j} - \hat{k}) \text{ N}$$

32.31. Model: Assume the magnetic field is uniform over the Hall probe.
Solve: The Hall voltage is given by Equation 32.24:

$$\Delta V_{\text{H}} = \frac{IB}{tne} \Rightarrow \frac{\Delta V_H}{B} = \frac{I}{tne}$$

In both uses, the quantities I, t, n, and e are unchanged, so the ratio $\dfrac{\Delta V_H}{B}$ is constant. Thus

$$\frac{1.9 \times 10^{-6} \text{ V}}{55 \times 10^{-3} \text{ T}} = \frac{2.8 \times 10^{-6} \text{ V}}{B} \Rightarrow B = 81 \text{ mT}$$

Section 32.8 Magnetic Forces on Current-Carrying Wires

32.33. Model: Assume that the field is uniform. The wire will float in the magnetic field if the magnetic force on the wire points upward and has a magnitude mg, allowing it to balance the downward gravitational force.
Solve: We can use the right-hand rule to determine which current direction experiences an upward force. The current being from right to left, the force will be *up* if the magnetic field $\vec{B}$ points out of the page. The forces will balance when

$$F = ILB = mg \Rightarrow B = \frac{mg}{IL} = \frac{(2.0 \times 10^{-3} \text{ kg})(9.8 \text{ m/s}^2)}{(1.5 \text{ A})(0.10 \text{ m})} = 0.131 \text{ T}$$

Thus $\vec{B} = (0.131 \text{ T, out of page})$.

Section 32.9 Forces and Torques on Current Loops

32.37. Solve: From Equation 32.28, the torque on the loop exerted by the magnetic field is

$$\vec{\tau} = \vec{\mu} \times \vec{B} \Rightarrow \tau = \mu B \sin \theta = IAB \sin \theta = (0.500 \text{ A})(0.050 \text{ m} \times 0.050 \text{ m})(1.2 \text{ T}) \sin 30° = 7.5 \times 10^{-4} \text{ N m}$$

32.39. Model: The torque on the current loop is due to the magnetic field produced by the current-carrying wire. Assume that the wire is very long.
Solve: (a) From Equation 32.27, the magnitude of the torque on the current loop is $\tau = \mu B \sin \theta$, where $\mu = I_{\text{loop}}$ A and B is the magnetic field produced by the current I_{wire} in the wire. The magnetic field of the wire is tangent to a circle around the wire. At the position of the loop, $\vec{B}$ points up and is $\theta = 90°$ from the axis of the loop. Thus,

$$\tau = (I_{\text{loop}} A) \frac{\mu_0 I_{\text{wire}}}{2\pi d} \sin \theta = \frac{(0.20 \text{ A})\pi (0.0010 \text{ m})^2 (2 \times 10^{-7} \text{ T m/A})(2.0 \text{ A}) \sin 90°}{2.0 \times 10^{-2} \text{ m}} = 1.26 \times 10^{-11} \text{ N m}$$

Note that the magnetic field produced by the wire on the current loop is *up* so that the angle θ between $\vec{B}$ and the normal to the loop is 90°.
(b) The loop is in equilibrium when $\theta = 0°$ or 180°. That is, when the coil is rotated by $\pm 90°$.

32.43. Model: Assume that the wires are infinitely long and that the magnetic field is due to currents in both the wires.

Visualize: Point 1 is a distance d_1 away from the two wires and point 2 is a distant d_2 away from the two wires. A right triangle with a 75° degree angle is formed by a straight line from point 1 to the intersection and a line from point 1 that is perpendicular to the wire. Likewise, point 2 makes a 15° right triangle.

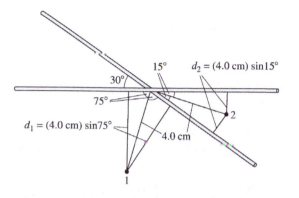

Solve: First we determine the distances d_1 and d_2 of the points from the two wires:

$$d_1 = (4.0 \text{ cm}) \sin 75° = 3.86 \text{ cm} = 0.0386 \text{ m}$$
$$d_2 = (4.0 \text{ cm}) \sin 15° = 1.04 \text{ cm} = 0.0104 \text{ m}$$

At point 1, the fields from both the wires point up and hence add. The total field is

$$B_1 = B_{\text{wire 1}} + B_{\text{wire 2}} = \frac{\mu_0}{2\pi} \frac{I_1}{d_1} + \frac{\mu_0}{2\pi} \frac{I_2}{d_1} = \frac{\mu_0}{\pi} \frac{(5.0 \text{ A})}{d_1} = \frac{(4 \times 10^{-7} \text{ T m/A})(5.0 \text{ A})}{0.0386 \text{ m}} = 5.2 \times 10^{-5} \text{ T}$$

In vector form, $\vec{B}_1 = (5.2 \times 10^{-5} \text{ T, out of page})$. Using the right-hand rule at point 2, the fields are in opposite directions but equal in magnitude. So, $\vec{B}_2 = \vec{0} \text{ T}$.

32.47. Model: Use the Biot-Savart law for a current-carrying segment.

Visualize: The distance from P to the inner arc is r_1 and the distance from P to the outer arc is r_2.

Solve: As given in Equation 32.6, the Biot-Savart law for a current-carrying small segment $\Delta \vec{s}$ is

$$\vec{B} = \frac{\mu_0}{4\pi} \frac{I \Delta \vec{s} \times \hat{r}}{r^2}$$

For the linear segments of the loop, $B_{\Delta s} = 0 \text{ T}$ because $\Delta \vec{s} \times \hat{r} = 0$. Consider a segment $\Delta \vec{s}$ on length on the inner arc. Because $\Delta \vec{s}$ is perpendicular to the $\hat{r}$ vector, we have

$$B = \frac{\mu_0}{4\pi} \frac{I \Delta s}{r_1^2} = \frac{\mu_0}{4\pi} \frac{I r_1 \Delta \theta}{r_1^2} = \frac{\mu_0}{4\pi} \frac{I \Delta \theta}{r_1} \Rightarrow B_{\text{arc 1}} = \int_{-\pi/2}^{\pi/2} \frac{\mu_0 I d\theta}{4\pi r_1} = \frac{\mu_0 I}{4\pi r_1} \pi = \frac{\mu_0 I}{4 r_1}$$

A similar expression applies for $B_{\text{arc 2}}$. The right-hand rule indicates an out-of-page direction for $B_{\text{arc 2}}$ and an into-page direction for $B_{\text{arc 1}}$. Thus,

$$\vec{B} = \left(\frac{\mu_0 I}{4 r_1}, \text{ into page} \right) + \left(\frac{\mu_0 I}{4 r_2}, \text{ out of page} \right) = \left[\frac{\mu_0 I}{4} \left(\frac{1}{r_1} - \frac{1}{r_2} \right), \text{ into page} \right]$$

The field strength is

$$B = \frac{(4\pi \times 10^{-7} \text{ T m/A})(5.0 \text{ A})}{4} \left(\frac{1}{0.010 \text{ m}} - \frac{1}{0.020 \text{ m}} \right) = 7.9 \times 10^{-5} \text{ T}$$

Thus $\vec{B} = (7.9 \times 10^{-5} \text{ T, into page})$.

32.49. Model: Assume that the solenoid is an ideal solenoid.

Solve: The magnetic field of a solenoid is $B_{sol} = \mu_0 NI/l$, where l is the length and N the total number of turns of wire. If the wires are wound as closely as possible, the spacing between one turn and the next is simply the diameter d_{wire} of the wire. The number of turns that will fit into length L is $N = l/d_{wire}$. For #18 wire, $N = (20\text{ cm})/(0.102\text{ cm}) = 196$ turns. The current required is

$$I_{\#18} = \frac{LB_{sol}}{\mu_0 N} = \frac{(0.20\text{ m})(5\times10^{-3}\text{ T})}{(4\pi\times10^{-7}\text{ T m/A})(196)} = 4.1\text{ A}$$

For #26 wire, $d_{wire} = 0.41$ mm, leading to $N = 488$ turns and $I_{\#26} = 1.63$ A. The current that would be needed with #26 wire exceeds the current limit of 1 A, but the current needed with #18 wire is within the current limit of 6 A. So use #18 wire with a current of 4.1 A.

32.53. Model: The heart is compared to a loop of current with radius 4.0 cm and magnetic field of 90 pT at its center.

Solve: (a) The current needed to produce this field can be computed from the equation for a magnetic field at the center of a loop.

$$I = \frac{2BR}{\mu_0} = \frac{2(90\times10^{-12}\text{ T})(0.040\text{ m})}{(4\pi\times10^{-7}\text{ T}\cdot\text{m/A})} = 5.73\times10^{-6}\text{ A} \approx 5.7\times10^{-6}\text{ A}$$

(b) $\mu = AI = \pi(0.040\text{ m})^2(5.73\times10^{-6}\text{ A}) = 2.88\times10^{-8}\text{ A m}^2 \approx 2.9\times10^{-8}\text{ A m}^2$

Assess: This is a small current, as we would have expected.

32.55. Model: The magnetic field is that of a current in the wire.

Solve: As given in Equation 32.6 for a current-carrying small segment $\Delta\vec{s}$, the Biot-Savart law is

$$\vec{B} = \frac{\mu_0}{4\pi}\frac{I\Delta\vec{s}\times\hat{r}}{r^2}$$

For the straight sections, $\Delta\vec{s}\times\hat{r} = 0$ because both $\Delta\vec{s}$ and $\hat{r}$ point along the same line. That is not the case with the curved section over which $\Delta\vec{s}$ and $\vec{r}$ are perpendicular. Thus,

$$B = \frac{\mu_0}{4\pi}\frac{I\Delta s}{r^2} = \frac{\mu_0}{4\pi}\frac{IR\,d\theta}{R^2} = \frac{\mu_0 I\,d\theta}{4\pi R}$$

where we used $\Delta s = R\Delta\theta \approx R\,d\theta$ for the small arc length Δs. Integrating to obtain the total magnetic field at the center of the semicircle,

$$B = \int_{-\pi/2}^{\pi/2} \frac{\mu_0 I\,d\theta}{4\pi R} = \frac{\mu_0 I}{4\pi R}\pi = \frac{\mu_0 I}{4R}$$

32.59. Model: Magnetic fields exert a force on moving charges.

Visualize:

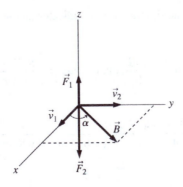

Solve: The right-hand rule applied to the first proton requires $\vec{B}$ lie toward the $+y$-axis from $\vec{v}_1$, while when applied to the second proton requires $\vec{B}$ lie toward the $+x$-axis from $\vec{v}_2$. Thus $\vec{B}$ lies in the first quadrant of the xy-plane. The force on each proton is

$$F_1 = qv_1 B \sin\alpha \qquad F_2 = qv_2 B \sin(90° - \alpha) = qv_1 B \cos\alpha$$

$$\Rightarrow \frac{F_1}{F_2} = \frac{v_1}{v_2}\tan\alpha \Rightarrow \alpha = \tan^{-1}\left(\frac{2.00\times10^6 \text{ m/s}}{1.00\times10^6 \text{ m/s}} \cdot \frac{1.20\times10^{-16} \text{ N}}{4.16\times10^{-16} \text{ N}}\right) = 30°$$

The magnetic field strength is thus

$$B = \frac{F_1}{qv_1\sin\alpha} = \frac{1.20\times10^{-16} \text{ N}}{(1.60\times10^{-19} \text{ C})(1.00\times10^6 \text{ m/s})\sin 30°} = 1.50\times10^{-3} \text{ T}$$

Thus $\vec{B} = (1.50 \text{ mT}, 30° \text{ ccw from } +x\text{-axis})$.

32.61. Model: Energy is conserved as the electron moves between the two electrodes. Assume the electron starts from rest. Once in the magnetic field, the electron moves along a circular arc.
Visualize:

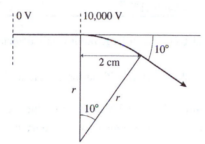

The electron is deflected by 10° after moving along a circular arc of angular width 10°.
Solve: Energy is conserved as the electron moves from the 0 V electrode to the 10,000 V electrode. The potential energy is $U = qV$ with $q = -e$, so

$$K_f + U_f = K_i + U_i \Rightarrow \tfrac{1}{2}mv^2 - eV = 0 + 0$$

$$v = \sqrt{\frac{2eV}{m}} = \sqrt{\frac{2(1.60\times10^{-19} \text{ C})(10,000 \text{ V})}{9.11\times10^{-31} \text{ kg}}} = 5.93\times10^7 \text{ m/s}$$

The radius of cyclotron motion in a magnetic field is $r = mv/eB$. From the figure we see that the radius of the circular arc is $r = (2.0 \text{ cm})/\sin 10°$. Thus

$$B = \frac{mv}{er} = \frac{(9.11\times10^{-31} \text{ kg})(5.93\times10^7 \text{ m/s})}{(1.60\times10^{-19} \text{ C})(0.020 \text{ m})/\sin 10°} = 2.9\times10^{-3} \text{ T}$$

32.65. Model: Charged particles moving perpendicular to a uniform magnetic field undergo circular motion at constant speed.
Solve: The potential difference causes an ion of mass m to accelerate from rest to a speed v. Upon entering the magnetic field, the ion follows a circular trajectory with cyclotron radius $r = mv/eB$. To be detected, an ion's trajectory must have radius $r = d/2 = 4.0000$ cm. This means the ion needs the speed

$$v = \frac{eBr}{m} = \frac{eBd}{2m}$$

This speed was acquired by accelerating from potential V to potential 0. We can use the conservation of energy equation to find the voltage that will accelerate the ion:

$$K_1 + U_1 = K_2 + U_2 \Rightarrow 0 \text{ J} + e\Delta V = \tfrac{1}{2}mv^2 + 0 \text{ J} \Rightarrow \Delta V = \frac{mv^2}{2e}$$

Using the above expression for v, the voltage that causes an ion to be detected is

$$\Delta V = \frac{mv^2}{2e} = \frac{m}{2e}\left(\frac{eBd}{2m}\right)^2 = \frac{eB^2d^2}{8m}$$

For example, the mass of N_2^+ is

$$m = m_N + m_N = 2(14.003\ u)(1.6605\times10^{-27}\ kg/u) = 4.6503\times10^{-26}\ kg$$

Note that we're given the atomic masses very accurately in Exercise 28. We need to retain this accuracy to tell the difference between N_2^+ and CO^+. The voltage for N_2^+ is

$$\Delta V = \frac{(1.6022\times10^{-19}\ C)(0.20000\ T)^2(0.080000\ m)^2}{8(4.6503\times10^{-26}\ kg)} = 110.07\ V$$

Ion	Mass(kg)	Accelerating voltage(V)
N_2^+	4.6503×10^{-26}	110.25
O_2^+	5.2969×10^{-26}	96.793
CO^+	4.6485×10^{-26}	110.29

Assess: The difference between N_2^+ and CO^+ is not large but is easily detectable.

32.67. Model: The loop will not rotate about the axle if the torque due to the magnetic force on the loop balances the torque of the weight.
Solve: The rotational equilibrium condition $\sum\vec{\tau}_{net} = 0\ N\,m$ is about the axle and means that the torque from the weight is equal and opposite to the torque from the magnetic force. We have

$$(50\times10^{-3}\ kg)g(0.025\ m) = \mu B\sin90° = (NIA)B$$

$$\Rightarrow B = \frac{(50\times10^{-3}\ kg)(9.8\ m/s^2)(0.025\ m)}{(10)(2.0\ A)(0.050\ m)(0.100\ m)} = 0.12\ T$$

Assess: The current in the loop must be clockwise for the two torques to be equal.

32.71. Model: A magnetic field exerts a magnetic force on a length of current-carrying wire.
Visualize:

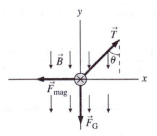

Solve: (a) The above figure shows a side view of the wire, with the current moving into the page. From the right-hand rule, the magnetic field $\vec{B}$ points *down* to give a leftward force on the current. The wire is hanging in static equilibrium, so $\vec{F}_{net} = \vec{F}_{mag} + \vec{F}_G + \vec{T} = 0\ N$. Consider a segment of wire of length L. The wire's linear mass density is $\mu = 0.050\ kg/m$, so the mass of this segment is $m = \mu L$ and its weight is $F_G = mg = \mu Lg$. The magnetic force on this length of wire is $F_{mag} = ILB$. In component form, Newton's first law is

$$(F_{net})_x = T\sin\theta - F_{mag} = T\sin\theta - ILB = 0\ N \Rightarrow T\sin\theta = ILB$$
$$(F_{net})_y = T\cos\theta - F_G = T\cos\theta - \mu Lg = 0\ N \Rightarrow T\cos\theta = \mu Lg$$

Dividing the first equation by the second,

$$\left[\frac{T\sin\theta}{T\cos\theta} = \tan\theta\right] = \left[\frac{ILB}{\mu Lg} = \frac{IB}{\mu g}\right] \Rightarrow B = \frac{\mu g\tan\theta}{I}$$

(b) $B = \dfrac{\mu g\tan\theta}{I} = \dfrac{(0.055 \text{ kg/m})(9.8 \text{ m/s}^2)\tan 11°}{10 \text{ A}} = 0.011456 \text{ T}$

The magnetic field is $\vec{B} = (11 \text{ mT, down})$.

32.73. Model: A current loop produces a magnetic field.
Visualize:

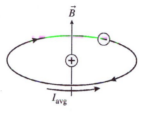

Solve: The field at the center of a current loop is $B_{\text{loop}} = \mu_0 I/2R$. The electron orbiting an atomic nucleus is, on average, a small current loop. Current is defined as $I = \Delta q/\Delta t$. During one orbital period T, the charge $\Delta q = e$ goes around the loop one time. Thus the *average* current is $I_{\text{avg}} = e/T$. For circular motion, the period is

$$T = \frac{2\pi R}{v} = \frac{2\pi(5.3\times10^{-11} \text{ m})}{2.2\times10^6 \text{ m/s}} = 1.514\times10^{-16} \text{ s} \Rightarrow I_{\text{avg}} = \frac{1.60\times10^{-19} \text{ C}}{1.514\times10^{-16} \text{ s}} = 1.057\times10^{-3} \text{ A}$$

Thus, the magnetic field at the center of the atom is

$$B_{\text{center}} = \frac{\mu_0 I_{\text{avg}}}{2R} = \frac{(4\pi\times10^{-7} \text{ T m/A})(1.057\times10^{-3} \text{ A})}{2(5.3\times10^{-11} \text{ m})} = 12.5 \text{ T} \approx 13 \text{ T}$$

ELECTROMAGNETIC INDUCTION

Exercises and Problems

Section 33.2 Motional emf

33.3. Visualize:

The wire is pulled at a constant speed in a magnetic field. This results in a motional emf and produces a current in the circuit. From energy conservation, the mechanical power provided by the puller must appear as electrical power in the circuit.

Solve: **(a)** Using Equation 33.6,

$$P = F_{pull}v \quad \Rightarrow \quad F_{pull} = \frac{P}{v} = \frac{4.0 \text{ W}}{4.0 \text{ m/s}} = 1.0 \text{ N}$$

(b) Using Equation 33.6 again,

$$P = \frac{v^2 l^2 B^2}{R} \quad \Rightarrow \quad B = \sqrt{\frac{RF_{pull}}{vl^2}} = \sqrt{\frac{(0.20 \text{ }\Omega)(1.0 \text{ N})}{(4.0 \text{ m/s})(0.10 \text{ m})^2}} = 2.2 \text{ T}$$

Assess: This is reasonable field for the circumstances given.

Section 33.3 Magnetic Flux

33.5. Model: Consider the solenoid to be long so the field is constant inside and zero outside.
Visualize: Please refer to Figure EX33.5. The field of a solenoid is along the axis. The flux through the loop is only nonzero inside the solenoid. Since the loop completely surrounds the solenoid, the total flux through the loop will be the same in both the perpendicular and tilted cases.
Solve: The field is constant inside the solenoid so we will use Equation 33.10. Take $\vec{A}$ to be in the same direction as the field. The magnetic flux is

$$\Phi = \vec{A}_{loop} \cdot \vec{B}_{loop} = \vec{A}_{sol} \cdot \vec{B}_{sol} = \pi r_{sol}^2 B_{sol} \cos\theta = \pi(0.010 \text{ m})^2(0.20 \text{ T}) = 6.3 \times 10^{-5} \text{ Wb}$$

When the loop is tilted the component of $\vec{B}$ in the direction of $\vec{A}$ is less, but the effective area of the loop surface through which the magnetic field lines cross is increased by the same factor, so the flux through the loop remains unchanged, as can be expected on physical grounds (i.e., we still have the same numbe of magnetic field lines going through the loop).

Section 33.4 Lenz's Law

33.9. Visualize: Please refer to Figure EX33.9. The changing current in the solenoid produces a changing flux in the loop. By Lenz's Law there will be an induced current and field to oppose the change in flux.
Solve: The current shown produces a field to the left inside the solenoid. So there is flux to the left through the surrounding loop. As the current in the solenoid increases there is more field and more flux to the left through the loop. There will be an induced current in the loop that will oppose the *change* by creating an induced field and flux to the right. By the right-hand rule, this requires a *clockwise* current.

Section 33.5 Faraday's Law

33.11. Model: Assume the field is uniform.
Visualize: Please refer to Figure EX33.11. If the changing field produces a changing flux in the loop there will be a corresponding induced emf and current.
Solve: (a) The induced emf is $\mathcal{E} = |d\Phi/dt|$ and the induced current is $I = \mathcal{E}/R$. The field B is changing, but the area A is not. Take $\vec{A}$ to be out of the page and parallel to $\vec{B}$, so $\Phi = AB$. Thus,

$$\mathcal{E} = \left| A\frac{dB}{dt} \right| = \left| \pi r^2 \frac{dB}{dt} \right| = \left| \pi(0.050 \text{ m})^2(0.50 \text{ T/s}) \right| = 3.9 \times 10^{-3} \text{ V}$$

$$I = \frac{\mathcal{E}}{R} = \frac{3.93 \times 10^{-3} \text{ V}}{0.20 \text{ }\Omega} = 2.0 \times 10^{-2} \text{ A} = 20 \text{ mA}$$

The field is increasing into the page. To prevent the increase, the induced field needs to point out of the page. Thus, the induced current must flow counterclockwise.
(b) As in part (a), $\mathcal{E} = A(dB/dt) = 3.9 \text{ mV}$ and $I = 20 \text{ mA}$. Here the field is out of the page and decreasing. To prevent the decrease, the induced field needs to point out of the page. Thus the induced current must flow counterclockwise.
(c) Now $\vec{A}$ (left or right) is perpendicular to $\vec{B}$ and so $\vec{A} \cdot \vec{B} = 0$ Wb. That is, the field does not penetrate the plane of the loop. If $\Phi = 0$ Wb, then $\mathcal{E} = |d\Phi/dt| = 0$ V/m and $I = 0$ A. There is *no* induced current.

Assess: Note that the induced field opposes the change.

33.13. Model: Assume the field strength is changing at a constant rate.
Visualize:

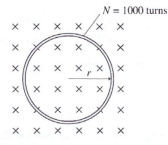

$N = 1000$ turns

The changing field produces a changing flux in the coil, and there will be a corresponding induced emf and current.
Solve: The induced emf of the coil is

$$\mathcal{E} = N\left| \frac{d\Phi}{dt} \right| = N\left| \frac{d(\vec{A} \cdot \vec{B})}{dt} \right| = NA\left| \frac{dB}{dt} \right| = N\pi r^2 \left| \frac{\Delta B}{\Delta t} \right| = (10^3)\pi(0.0050 \text{ m})^2 \left(\frac{0.20 \text{ T}}{10 \times 10^{-3} \text{ s}} \right) = 1.6 \text{ V}$$

where we've used the fact that $\vec{B}$ is parallel to $\vec{A}$.

Assess: This seems to be a reasonable emf as there are many turns. Notice that the sign of the change in B does not enter the calculation.

Section 33.6 Induced Fields

33.17. Model: Assume the magnetic field inside the circular region is uniform. The region looks exactly like a cross section of a solenoid of radius 2.0 cm. The proton accelerates due to the force of the induced electric field.

Visualize: Please refer to Figure EX33.17.

Solve: Equation 33.26 gives the strength of the induced electric field inside a solenoid with changing magnetic field. The magnitude at each point is thus $E = \frac{r}{2}\left|\frac{dB}{dt}\right|$, where r indicates the radial distance between each point and the center of the circular region. The direction of E can be determined with Lenz's Law. As B decreases, a clockwise electric field is induced that would create a current that opposes the decrease if a conducting loop were present. Using Newton's second law,

$$F = eE = ma \quad \Rightarrow \quad a = \frac{eE}{m} = \frac{er}{2m}\left|\frac{dB}{dt}\right|$$

The sign indicates that the acceleration is in the same direction as the induced electric field. At point a the acceleration is

$$a = \frac{-(1.6\times10^{-19}\text{ C})(1.0\times10^{-2}\text{ m})}{2(1.6\times10^{-27}\text{ kg})}(0.10\text{ T/s}) \quad \Rightarrow \quad \vec{a} = (4.8\times10^4\text{ m/s}^2,\text{ up})$$

At point b, $r = 0$ m so $a = 0$ m/s^2. At point c, $\vec{a} = (4.8\times10^4\text{ m/s}^2,\text{ down})$. At point d, $\vec{a} = (9.6\times10^4\text{ m/s}^2,\text{ down})$.

Section 33.8 Inductors

33.19. Model: Assume that the current changes uniformly.

Visualize: We want to increase the current without exceeding the maximum potential difference.

Solve: Since we want the minimum time, we will use the maximum potential difference:

$$\Delta V = -L\frac{dI}{dt} = -L\frac{\Delta I}{\Delta t} \quad \Rightarrow \quad \Delta t = L\left|\frac{\Delta I}{\Delta V_{max}}\right| = (200\times10^{-3}\text{ H})\frac{3.0\text{ A} - 1.0\text{ A}}{400\text{ V}} = 1.0\text{ ms}$$

Assess: If we change the current in over a shorter time the potential difference will exceed the limit.

Section 33.9 *LC* Circuits

33.23. Model: Changing the variable capacitor in combination with a fixed inductor will change the resonant frequency of the *LC* circuit.

Solve: Because the resonant frequency depends on the inverse square root of the capacitance, a lower capacitance will produce a higher frequency and vice versa. The maximum frequency is

$$\omega_{max} = \sqrt{\frac{1}{LC_{min}}} = \sqrt{\frac{1}{(2.0\times10^{-3}\text{ H})(100\times10^{-12}\text{ F})}} = 2.2\times10^6\text{ rad/s}$$

The corresponding minimum is $\omega_{min} = 2.0\times10^6$ rad/s. These are angular frequencies so we can use $f = \omega/(2\pi)$ to find $f_{min} = 2.5\times10^5$ Hz and $f_{max} = 3.6\times10^5$ Hz, giving a range of 250 to 360 kHz.

Section 33.10 *LR* Circuits

33.25. Visualize: Please refer to Figure EX33.25. This is a simple *LR* circuit if the resistors in parallel are treated as an equivalent resistor in series with the inductor.

Solve: We can find the equivalent resistance from the time constant since we know the inductance. We have

$$\tau = \frac{L}{R_{eq}} \quad \Rightarrow \quad R_{eq} = \frac{L}{\tau} = \frac{7.5 \times 10^{-3}\ \text{H}}{25 \times 10^{-6}\ \text{s}} = 300\ \Omega$$

The equivalent resistance is the parallel addition of the unknown resistor R and $500\ \Omega$. We have

$$\frac{1}{R_{eq}} = \frac{1}{500\ \Omega} + \frac{1}{R} \quad \Rightarrow \quad R = \frac{(500\ \Omega)(300\ \Omega)}{500\ \Omega - 300\ \Omega} = 750\ \Omega$$

33.29. Model: Assume the field is uniform in space although it is changing in time.
Visualize: The changing magnetic field strength produces a changing flux through the loop, and a corresponding induced emf and current.
Solve: Since the field is perpendicular to the plane of the loop, $\vec{A}$ is parallel to $\vec{B}$ and $\Phi = AB$. The emf is

$$\varepsilon = \left| \frac{d\Phi}{dt} \right| = A \left| \frac{dB}{dt} \right| = (0.20\ \text{m})^2 \left| (4 - 4t)\text{T/s} \right| = \left| 0.16(1 - t) \right|\ \text{V} \quad \Rightarrow \quad I = \frac{\varepsilon}{R} = \left| 1.6(1 - t) \right|\ \text{A}$$

The magnetic field is increasing over the interval $0\ \text{s} < t < 1\ \text{s}$ and is decreasing over the interval $1\ \text{s} < t < 2\ \text{s}$, so the induced emf and current must have opposite signs in the second half of the time interval. We arbitrarily choose the sign to be positive during the first half. Inserting $t = 0.0$, 1.0, and $2.0\ \text{s}$ into the equation above gives a current of 1.6, 0.0, and -1.6 A, respectively.

33.31. Model: Assume that the field is uniform in space over the coil.
Visualize: We want an induced current so there must be an induced emf created by a changing flux.
Solve: To relate the emf and the current we need to know the resistance, which may be found using Equation 30.22 and the resistivity values from Table 30.2:

$$R = \frac{\rho_{Cu} l_{wire}}{A_{wire}} = \frac{\rho_{Cu} N 2\pi r}{\pi r_{wire}^2} = \frac{2(100)(1.7 \times 10^{-8}\ \Omega\ \text{m})(0.040\ \text{m})}{(0.25 \times 10^{-3}\ \text{m})^2} = 2.176\ \Omega$$

The magnetic field is parallel to the axis of the coil so the flux for a single loop of the coil is $\Phi = \vec{A} \cdot \vec{B} = BA$, if we take the normal to a coil loop to be in the same direction as the field. Using Faraday's law,

$$\varepsilon = IR = NA \left| \frac{dB}{dt} \right| = N\pi r^2 \left| \frac{dB}{dt} \right| \quad \Rightarrow \quad \left| \frac{dB}{dt} \right| = \frac{IR}{N\pi r^2} = \frac{(2.0\ \text{A})(2.176\ \Omega)}{100\pi(0.040\ \text{m})^2} = 8.7\ \text{T/s}$$

33.35. Model: Assume the wire is long enough so we can use the formula for the magnetic field of an "infinite" wire.
Visualize:

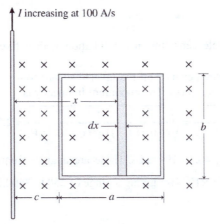

The magnetic field in the vicinity of the loop is due to the current in the wire and is perpendicular to the loop. The current is changing so the field and the flux through the loop are changing. This will create an induced emf and induced current in the loop.
Solve: The induced current depends on the induced emf and is

$$I_{\text{loop}} = \frac{\mathcal{E}_{\text{loop}}}{R} = \frac{1}{R}\left|\frac{d\Phi}{dt}\right|$$

The flux through a rectangular loop due to a wire was found in Example 33.5:

$$\Phi = \frac{\mu_0 I b}{2\pi}\ln\left(\frac{c+a}{c}\right)$$

We have $c = 0.10$ m and $a = b = 0.20$ m. The current through the loop is therefore

$$I_{\text{loop}} = \frac{1}{R}\frac{\mu_0 b}{2\pi}\ln\left(\frac{c+a}{c}\right)\frac{dI}{dt} = \frac{(4\pi\times10^{-7}\text{ T m/A})(0.020\text{ m})}{(0.010\,\Omega)2\pi}\ln\left(\frac{0.030\text{ m}}{0.010\text{ m}}\right)(100\text{ A/s}) = 44\ \mu\text{A}$$

33.37. Model: Since the solenoid is fairly long compared to its diameter and the loop is located near the center, assume the solenoid field is uniform inside and zero outside.
Visualize: Please refer to Figure P33.37. The solenoid's magnetic field is perpendicular to the loop and creates a flux through the loop. This flux changes as the solenoid's current changes, causing an induced emf and corresponding induced current.
Solve: (a) Using Faraday's law, the induced current is

$$I_{\text{loop}} = \frac{\mathcal{E}_{\text{loop}}}{R} = \frac{1}{R}\left|\frac{d\Phi}{dt}\right| = \frac{A_{\text{sol}}}{R}\left|\frac{dB_{\text{sol}}}{dt}\right| = \frac{\pi r_{\text{sol}}^2}{R}\left|\frac{d}{dt}\left(\frac{\mu_0 NI}{l}\right)\right| = \frac{\pi r_{\text{sol}}^2\mu_0 N}{Rl}\left|\frac{dI}{dt}\right|$$

where we have used the fact that the field is approximately zero outside the solenoid, so the flux is confined to the area $A_{\text{sol}} = \pi r_{\text{sol}}^2$ of the solenoid, *not* the larger area of the loop itself. The current is constant from $0\text{ s} \le t \le 1\text{ s}$ so $dI/dt = 0$ A/s and $I_{\text{loop}} = 0$ A during this interval. Thus, $I(0.5\text{s}) = 0.0$ A.

(b) During the interval $1\text{ s} \le t \le 2\text{ s}$, the current is changing at the rate $|dI/dt| = 40$ A/s, so the current during this interval is

$$I_{\text{loop}} = \frac{\pi(0.010\text{ m})^2(4\pi\times10^{-7}\text{ T m/A})(100)}{(0.10\,\Omega)(0.10\text{ m})}(40\text{ A/s}) = 1.58\times10^{-4}\text{ A} = 158\ \mu\text{A}$$

Thus $I(1.5\text{ s}) = 0.16$ mA.

(c) The current is constant from $2\text{ s} \le t \le 3\text{ s}$, so again $dI/dt = 0$ A/s and $I_{\text{loop}} = 0$ A so $I(2.5\text{s}) = 0.0$ A.

Assess: The induced current is proportional to the negative derivative of the solenoid current, and is zero when there is a constant current in the solenoid.

33.39. Model: Assume that the magnetic field of coil 1 passes entirely through coil 2 and that we can use the magnetic field of a solenoid for coil 1.
Visualize: Please refer to Figure P33.39. The field of coil 1 produces flux in coil 2. The changing current in coil 1 gives a changing flux in coil 2 and a corresponding induced emf and current in coil 2.
Solve: (a) From 0 s to 0.1 s and 0.3 s to 0.4 s the current in coil 1 is constant, so its magnetic field is constant and the current in coil 2 must be zero. Thus $I(0.05\text{ s}) = 0$ A.

(b) From 0.1 s to 0.3 s, the induced current from the induced emf is given by Faraday's law. The current in coil 2 is

$$I_2 = \frac{\mathcal{E}_2}{R} = \frac{1}{R}N_2\left|\frac{d\Phi_2}{dt}\right| = \frac{1}{R}N_2 A_2\left|\frac{dB}{dt}\right| = \frac{1}{R}N_2\pi r_2^2\left|\frac{d}{dt}\left(\frac{\mu_0 N_1 I_1}{l_1}\right)\right| = \frac{N_2\pi r_2^2\mu_0 N_1}{Rl_1}\left|\frac{dI_1}{dt}\right|$$

$$= \frac{20\pi(0.010\text{ m})^2(4\pi\times10^{-7}\text{ T m/A})(20)}{(2\,\Omega)(20)(0.0010\text{ m})}|20\text{ A/s}| = 7.95\times10^{-5}\text{ A} = 79\ \mu\text{A}$$

We used the facts that the field of coil 1 is constant inside the loops of coil 2 and the flux is confined to the area $A_2 = \pi r_2^2$ of coil 2. Also, we used $l_1 = N_1 d = 20(1.0\text{ mm})$ and $|dI/dt| = (4\text{ A})/(0.2\text{ s}) = 20$ A/s. From 0.1 s to 0.2 s the current in coil 1 is initially negative so the field is initially to the right and the flux is decreasing. The induced current will *oppose this change* and will therefore produce a field to the right. This requires an induced current in coil 2 that comes out of the page at the top of the loops so it is negative. From 0.2 s to 0.3 s the current in coil 1 is positive so the field is to the left and the flux is increasing. The induced current will *oppose this change* and will therefore produce a field to the right. Again, this is a negative current. Hence $I(0.25\text{ s}) = 79\ \mu\text{A}$ right to left through the resistor.

33.45. Model: Assume the field changes abruptly at the boundary and is uniform.

Visualize: Please refer to Figure P33.45. As the loop enters the field region the amount of flux will change as more area has field penetrating it. This change in flux will create an induced emf and corresponding current. While the loop is moving at constant speed, the rate of change of the area is not constant because of the orientation of the loop. The loop is moving along the x-axis.

Solve: (a) If the edge of the loop enters the field region at $t = 0$ s, then the leading corner has moved a distance $x = v_0 t$ at time t. The area of the loop with flux through it is

$$A = 2\left(\tfrac{1}{2}\right)yx = x^2 = (v_0 t)^2$$

where we have used the fact that $y = x$ since the sides of the loop are oriented at $45°$ to the horizontal. Take the surface-normal of the loop to be into the page so that $\Phi = \vec{A} \cdot \vec{B} = BA$. The current in the loop is

$$I = \frac{\mathcal{E}}{R} = \frac{1}{R}\left|\frac{d\Phi}{dt}\right| = \frac{1}{R}B\left|\frac{dA}{dt}\right| = \frac{1}{R}B\left|\frac{d}{dt}(v_0 t)^2\right| = \frac{1}{R}B(2)v_0^2 t = \left(\frac{2(0.80 \text{ T})(10 \text{ m/s})^2}{0.10 \,\Omega}\right)t = (1.6\times10^3 \text{ A})t$$

The current is increasing at a constant rate. This expression is good until the loop is halfway into the field region. The time for the loop to be halfway is found as follows:

$$\frac{10 \text{ cm}}{\sqrt{2}} = v_0 t = (10 \text{ m/s})t \quad \Rightarrow \quad t = 7.1\times10^{-3} \text{ s} = 7.1 \text{ ms}$$

At this time the current is 11 A. While the second half of the loop is moving into the field, the flux continues to increase, but at a slower rate. Therefore, the current will decrease at the same rate as it increased before, until the loop is completely in the field at $t = 14$ ms. After that the flux will not change and the current will be zero.

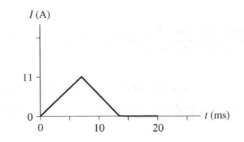

(b) The maximum current of 11 A occurs when the flux is changing the fastest, and this occurs when the loop is halfway into the region of the field.

33.47. Visualize: Please refer to Figure 33.26. The moving wire in a magnetic field results in a motional emf and a current in the loop.

Solve: (a) The induced emf is

$$\mathcal{E}_{\text{loop}} = vlB = (100 \text{ m/s})(0.040 \text{ m})(1.0 \text{ T}) = 4.0 \text{ V}$$

(b) The total resistance of the square loop is $R = 4(0.010 \,\Omega) = 0.040 \,\Omega$. The induced current is

$$I = \mathcal{E}_{\text{loop}}/R = (4.0 \text{ V})/(0.040 \,\Omega) = 100 \text{ A}$$

(c) The slide wire generates the emf, $\mathcal{E}_{\text{loop}}$. However, the slide wire also has an "internal resistance" due to the resistance of the slide wire itself—namely $r = 0.01\,\Omega$. Thus, the slide wire acts like a battery with an internal resistance. The potential difference between the ends is

$$\Delta V = \mathcal{E}_{\text{loop}} - Ir = 4.0 \text{ V} - (100 \text{ A})(0.01 \,\Omega) = 3.0 \text{ V}$$

It is this 3.0 V that drives the 100 A current through the remaining $0.03 \,\Omega$ resistance of the "external circuit."

Assess: The slide wire system acts as an emf just like a battery does.

33.49. Model: Assume that the magnetic field is uniform in the region of the loop.

Visualize: Please refer to Figure P33.49. The rotating semicircle will change the area of the loop and therefore the flux through the loop. This changing flux will produce an induced emf and corresponding current in the bulb.

Solve: (a) The spinning semicircle has a normal to the surface that changes in time, so while the magnetic field is constant, the area is changing. The flux through in the lower portion of the circuit does not change and will not contribute to the emf. Only the flux in the part of the loop containing the rotating semicircle will change. The flux associated with the semicircle is

$$\Phi = \vec{A} \cdot \vec{B} = BA = BA\cos\theta = BA\cos(2\pi f t)$$

where $\theta - 2\pi f t$ is the angle between the normal of the rotating semicircle and the magnetic field and A is the area of the semicircle. The induced current from the induced emf is given by Faraday's law. We have

$$I = \frac{\mathcal{E}}{R} = \frac{1}{R}\left|\frac{d\Phi}{dt}\right| = \frac{1}{R}\left|\frac{d}{dt}BA\cos(2\pi f t)\right| = \frac{B}{R}\frac{\pi r^2}{2}2\pi f\sin(2\pi f t)$$

$$= \frac{2(0.20\ \text{T})\pi^2(0.050\ \text{m})^2}{2(1.0\ \Omega)}f\sin(2\pi f t) = (4.9\times10^{-3})f\sin(2\pi f t)\ \text{A}$$

where the frequency f is in Hz.

(b) We can now solve for the frequency necessary to achieve a certain current. From our study of DC circuits we know how power relates to resistance:

$$P = I^2 R \quad \Rightarrow \quad I = \sqrt{P/R} = \sqrt{(4.0\ \text{W})/(1.0\ \Omega)} = 2.0\ \text{A}$$

The maximum of the sine function is +1, so the maximum current is

$$I_{max} = 4.9\times10^{-3}f \quad \text{A s} = 2.0\ \text{A} \quad \Rightarrow \quad f = \frac{2.0\ \text{A}}{4.9\times10^{-3}\ \text{A s}} = 4.1\times10^2\ \text{Hz}$$

Assess: This is not a reasonable frequency to obtain by hand.

33.53. Model: Assume the magnetic field is uniform in the region of the loop.

Visualize:

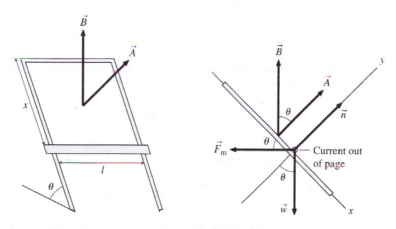

The moving wire creates a changing area and corresponding change in flux. This produces an induced emf and induced current. The flux through the loop depends on the size and orientation of the loop.

Solve: (a) The normal to the surface is perpendicular to the loop and the flux is $\Phi_{inner} = \vec{A} \cdot \vec{B} = AB\cos\theta$. We can get the current from Faraday's law. Since the loop area is $A = lx$, We have

$$I = \frac{\mathcal{E}}{R} = \frac{1}{R}\left|\frac{d\Phi}{dt}\right| = \frac{1}{R}\left|\frac{d}{dt}lxB\cos\theta\right| = \frac{Bl\cos\theta}{R}\left|\frac{dx}{dt}\right| = \frac{Blv\cos\theta}{R}$$

(b) Using the free-body diagram shown in the figure, we can apply Newton's second law. The magnetic force on a straight, current-carrying wire is $F_m = IlB$ and is horizontal. Using the current I from part (a) gives

$$\sum F_x = -F_m \cos\theta + mg\sin\theta = -\frac{B^2 l^2 v \cos^2\theta}{R} + mg\sin\theta = ma_x$$

Terminal speed is reached when a_x drops to zero. In this case, the two terms are equal and we have

$$v_{term} = \frac{mgR\tan\theta}{l^2 B^2 \cos\theta}$$

33.55. Model: Assume the magnetic field is uniform over the loop.
Visualize: The motion of the eye will change the orientation of the loop relative to the fixed field direction, resulting in an induced emf.
Solve: We are just interested in the emf due to the motion of the eye, so we can ignore the details of the time dependence of the change. From Faraday's law,

$$\mathcal{E}_{coil} = N\left|\frac{d\Phi}{dt}\right| = N\left|\frac{\Delta\Phi}{\Delta t}\right| = NAB\left|\frac{\Delta\cos\theta}{\Delta t}\right| = \frac{N\pi r^2 B|\cos\theta_f - \cos\theta_i|}{\Delta t}$$

$$= \frac{20\pi(3.0\times10^{-3}\text{ m})^2(1.0\text{ T})|\cos(85°) - \cos(90°)|}{0.20\text{ s}} = 2.5\times10^{-4}\text{ V}$$

Assess: This is a reasonable emf to measure, although you might need some amplification.

33.57. Model: Assume the field is uniform in the region of the coil.
Visualize:

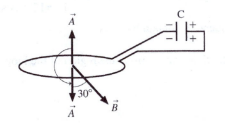

The rotation of the coil in the field will change the flux and produce an induced emf and a corresponding induced current. The current will charge the capacitor.
Solve: The induced current is

$$I_{coil} = \frac{\mathcal{E}_{coil}}{R} = \frac{N}{R}\left|\frac{d\Phi}{dt}\right|$$

The definition of current is $I = dq/dt$. Consequently, the charge flow through the coil and onto the capacitor is given by

$$\frac{dq}{dt} = \frac{N}{R}\left|\frac{d\Phi}{dt}\right| \quad\Rightarrow\quad \frac{\Delta q}{\Delta t} = \frac{N}{R}\left|\frac{\Delta\Phi}{\Delta t}\right| \quad\Rightarrow\quad \Delta q = \frac{N}{R}|\Delta\Phi| = \frac{N}{R}(\Phi_f - \Phi_i)$$

We are only interested in the total charge that flows due to the change in flux and not the details of the time dependence. In this case, the flux is changed by physically rotating the coil in the field. The flux is $\Phi = \vec{A}\cdot\vec{B} = AB\cos\theta$. The change in flux is

$$\Delta\Phi = AB(\cos\theta_f - \cos\theta_i) = \pi(0.020\text{ m})^2(55\times10^{-6}\text{ T})(\cos30° - \cos210°) = 1.2\times10^{-7}\text{ Wb}$$

Note that the field is 60° from the horizontal and the normal to the plane of the loop is vertical. The final angle, when $\vec{A}$ points down, is $\theta_f = 30°$, so the initial angle is $\theta_i = \theta_f + 180° = 210°$. The charge that flows onto the capacitor is

$$\Delta q = \frac{N}{R}(\Phi_f - \Phi_i) = \frac{200(1.2\times10^{-7}\text{ Wb})}{2.0\text{ }\Omega} = 1.2\times10^{-5}\text{ C} \quad\Rightarrow\quad \Delta V_C = \frac{\Delta q}{C} = \frac{1.2\times10^{-5}\text{ C}}{1.0\times10^{-6}\text{ F}} = 12\text{ V}$$

33.61. Model: Assume the straight wire in part (b) is "infinite."
Visualize: The energy density depends on the field strength in a particular region of space.
Solve: (a) We will use the formula for the magnetic field at the center of a current loop and then find the energy density. We have

$$B = \frac{\mu_0 I}{2 r_{\text{loop}}} = \frac{(4\pi \times 10^{-7} \text{ T m/A})(1.0 \text{ A})}{2(0.020 \text{ m})} = 3.14 \times 10^{-5} \text{ T}$$

$$u_B = \frac{1}{2\mu_0} B^2 = \frac{(3.14 \times 10^{-5} \text{ T})^2}{2(4\pi \times 10^{-7} \text{ T m/A})} = 3.9 \times 10^{-4} \text{ J/m}^3$$

(b) We will use the standard formula for the magnetic field of a long straight wire. We have

$$B = \frac{\mu_0 I}{2\pi r} \quad \Rightarrow \quad u_B = \frac{1}{2\mu_0} B^2 = \frac{1}{2\mu_0} \left(\frac{\mu_0 I}{2\pi r} \right)^2 \quad \Rightarrow \quad I = \sqrt{\frac{8\pi^2 r^2 u_B}{\mu_0}} = \sqrt{\frac{8\pi^2 (0.020 \text{ m})^2 (3.9 \times 10^{-4} \text{ J/m}^3)}{(4\pi \times 10^{-7} \text{ T m/A})}} = 3.1 \text{ A}$$

33.63. Model: Assume the field is constant over the cross section of the patient.
Visualize: The changing field results in a changing flux and an induced emf.
Solve: To determine the largest emf, assume the normal to the surface is parallel to the field. From Faraday's law, the emf is

$$\mathcal{E} = \left| \frac{d\Phi}{dt} \right| = \left| \frac{d}{dt} AB \right| = A \left| \frac{dB}{dt} \right| = A \left| \frac{\Delta B}{\Delta t} \right|$$

The time interval acceptable for the change is

$$\Delta t = A \frac{|\Delta B|}{\mathcal{E}} = \frac{(0.060 \text{ m}^2)|5.0 \text{ T} - 0 \text{ T}|}{0.10 \text{ V}} = 3.0 \text{ s}$$

Assess: This is a reasonable amount of time in which to change the field.

33.67. Visualize: The current through the inductor is changing with time, which leads to a changing potential difference across the inductor.
Solve: (a) To find the potential difference we differentiate the current. We have

$$\Delta V_L = -L \frac{dI}{dt} = -L \frac{d}{dt} (I_0 e^{-t/\tau}) = \left(\frac{L I_0}{\tau} \right) e^{-t/\tau}$$

(b) For $t = 1$ ms, the potential difference is

$$\Delta V_L = \frac{(20 \times 10^{-3} \text{ H})(50 \times 10^{-3} \text{ A})}{1.0 \times 10^{-3} \text{ s}} e^{-(1.0 \times 10^{-3} \text{ s})/(1.0 \times 10^{-3} \text{ s})} = (1.0 \text{ V}) e^{-1} = 0.37 \text{ V}$$

Similar calculations give 1.0 V, 0.13 V, and 0.05 V for $t = 0$ ms, 2 ms, and 3 ms.

33.69. Model: Assume any resistance is negligible.
Visualize: The potential difference across the inductor and capacitor oscillate. The period of oscillation depends on the resistance and capacitance, and the potential difference across the capacitor depends on the charge.
Solve: The current is $I(t) = I_0 \cos \omega t = (0.60 \text{ A}) \cos \omega t$. We can relate the extreme values of the current and the capacitor potential difference. Using $\omega = 1/\sqrt{LC}$ and $Q = C\Delta V_C$, we find

$$I_0 = \omega Q_0 = \omega C \Delta V_{C, \text{max}} = \frac{1}{\sqrt{LC}} C \Delta V_{C, \text{max}} \quad \Rightarrow \quad C = \frac{I_0^2 L}{\Delta V_{C, \text{max}}^2} = \frac{(0.60 \text{ A})^2 (10 \times 10^{-3} \text{ H})}{(60 \text{ V})^2} = 1.0 \ \mu\text{F}$$

Assess: This is a reasonable capacitance.

33.73. Model: Assume negligible resistance in the LC part of the circuit.

Visualize: With the switch in position 1 for a long time the capacitor is fully charged. After moving the switch to position 2, there will be oscillations in the LC part of the circuit.

Solve: (a) After a long time the potential across the capacitor will be that of the battery and $Q_0 = C\Delta V_{\text{batt}}$. When the switch is moved, the capacitor will discharge through the inductor and LC oscillations will begin. The maximum current is

$$I_0 = \omega Q_0 = \omega C \Delta V_{\text{batt}} = \frac{C \Delta V_{\text{batt}}}{\sqrt{LC}} = \sqrt{\frac{C}{L}} \Delta V_{\text{batt}} = \sqrt{\frac{(2.0 \times 10^{-6} \text{ F})}{(50 \times 10^{-3} \text{ H})}} (12 \text{ V}) = 7.6 \times 10^{-2} \text{ A} = 76 \text{ mA}$$

(b) The current will be a maximum one-quarter cycle after the maximum charge. The period is

$$T = \frac{2\pi}{\omega} = 2\pi\sqrt{LC} = 2\pi\sqrt{(50 \times 10^{-3} \text{ H})(2.0 \times 10^{-6} \text{ F})} = 2.0 \text{ ms}$$

So the current is first maximum at $t_{\text{max}} = \frac{1}{4}T = 0.50$ ms.

33.77. Visualize: Please refer to Figure P33.77.

Solve: (a) After a long time has passed the current will no longer be changing. With steady currents, the potential difference across the inductor is $\Delta V_L = -L(dI/dt) = 0$ V. An ideal inductor has no resistance ($R = 0\,\Omega$), so the inductor simply acts like a wire. The circuit is simply that of a battery and resistor R, so the current is $I_0 = \Delta V_{\text{bat}}/R$.

(b) In general, we need to apply Kirchhoff's loop law to the circuit. Starting with the battery and going clockwise, the loop law is

$$\Delta V_{\text{bat}} + \Delta V_R + \Delta V_L = \Delta V_{\text{bat}} - IR - L\frac{dI}{dt} = 0$$

$$\frac{dI}{dt} = \frac{\Delta V_{\text{bat}}}{L} - \frac{IR}{L} = \frac{R}{L}\left(\frac{\Delta V}{R} - I\right) = \frac{R}{L}(I_0 - I)$$

This is a differential equation that we can solve by direct integration. The current is $I = 0$ A at $t = 0$ s, so separate the current and time variables and then integrate from 0 A at 0 s to current I at time t:

$$\int_0^I \frac{dI}{I_0 - I} = \frac{R}{L}\int_0^t dt \quad \Rightarrow \quad -\ln(I_0 - I)\big|_0^I = \frac{R}{L}t\big|_0^t \quad \Rightarrow \quad -\ln\left(\frac{I_0 - I}{I_0}\right) = \frac{t}{L/R}$$

Taking the exponential of both sides gives

$$\frac{I_0 - I}{I_0} = e^{-t/(L/R)} \quad \Rightarrow \quad I_0 - I = I_0 e^{-t/(L/R)}$$

Finally, solving for I gives

$$I = I_0(1 - e^{-t/(L/R)})$$

The current is 0 A at $t = 0$ s, as expected, and exponentially approaches I_0 as $t \to \infty$.

(c)

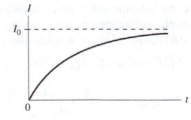

Assess: The current is zero at the start and approaches the steady final value. The behavior is similar to the charging of a capacitor.

ELECTROMAGNETIC FIELDS AND WAVES

Exercises and Problems

Section 34.1 *E* or *B*? It Depends on Your Perspective

34.1. **Model:** Apply the Galilean transformation of velocity.
Solve: (a) In the laboratory frame, the speed of the proton is

$$v = \sqrt{(1.41\times10^6 \text{ m/s})^2 + (1.41\times10^6 \text{ m/s})^2} = 2.0\times10^6 \text{ m/s}$$

The angle the velocity vector makes with the positive *y*-axis is

$$\theta = \tan^{-1}\left(\frac{1.41\times10^6 \text{ m/s}}{1.41\times10^6 \text{ m/s}}\right) = 45°$$

(b) In the rocket frame, we need to first determine the vector $\vec{v}'$. Equation 34.1 yields:

$$\vec{v}_{PR} = \vec{v}_{PL} + \vec{v}_{LR} = (1.41\times10^6\hat{i} + 1.41\times10^6\hat{j}) \text{ m/s} + (-1.00\times10^6\hat{i}) \text{ m/s} = (0.41\times10^6\hat{i} + 1.41\times10^6\hat{j}) \text{ m/s}$$

The speed of the proton is

$$v_{PL} = \sqrt{(0.41\times10^6 \text{ m/s})^2 + (1.41\times10^6 \text{ m/s})^2} = 1.47\times10^6 \text{ m/s}$$

The angle the velocity vector makes with the positive *y*-axis is

$$\theta' = \tan^{-1}\left(\frac{0.41\times10^6 \text{ m/s}}{1.41\times10^6 \text{ m/s}}\right) = 16.2°$$

34.5. **Model:** Use the Galilean transformation of fields.
Visualize: We are given $\vec{v}_{BA} = 1.0\times10^6\hat{i}$ m/s, $\vec{B}_A = 0.50\hat{k}$ T, and $\vec{E}_A = \left(\frac{1}{\sqrt{2}}\hat{i} + \frac{1}{\sqrt{2}}\hat{j}\right)\times10^6$ V/m.

Solve: Equation 34.10 gives the Galilean transformation equation for the electric field in different frames:

$$\vec{E}_B = \vec{E}_A + \vec{v}_{BA} \times \vec{B}_A \qquad\qquad \vec{B}_B = \vec{B}_A - \frac{1}{c^2}\vec{v}_{BA} \times \vec{E}_A$$

The electric field from the moving rocket is

$$\vec{E}' = (\hat{i} + \hat{j})0.707\times10^6 \text{ V/m} + (1.0\times10^6\hat{i} \text{ m/s})\times(0.50\hat{k} \text{ T}) = (0.707\times10^6\hat{i} + 0.207\times10^6\hat{j}) \text{ V/m}$$

$$\theta = \tan^{-1}\left(\frac{0.207\times10^6 \text{ V/m}}{0.707\times10^6 \text{ V/m}}\right) = 16.3° \text{ above the } x\text{-axis}$$

Section 34.2 The Field Laws Thus Far

Section 34.3 The Displacement Current

34.7. Solve: The units of $\varepsilon_0 (d\Phi_e/dt)$ are

$$\frac{C^2}{N\,m^2} \times \frac{(N/C)(m^2)}{s} = \frac{C}{s} = A$$

34.9. Model: Use the results of Exercise EX34.8.
Solve:

$$I_{disp} = C\frac{dV_C}{dt} \quad \Rightarrow \quad C = \frac{I_{disp}}{\dfrac{dV_C}{dt}} = \frac{1.0\,A}{1.0\times10^6\,V/s} = 1.0\,\mu F$$

Assess: This is a typical capacitance.

Section 34.5 Electromagnetic Waves

34.13. Model: The electric and magnetic field amplitudes of an electromagnetic wave are related.
Solve: Using Equation 34.29,

$$B_0 = \frac{E_0}{c} = \frac{10\,V/m}{3\times10^8\,m/s} = 3.3\times10^{-8}\,Vs/m^2 = 3.3\times10^{-8}\,T$$

Section 34.6 Properties of Electromagnetic Waves

34.17. Model: The electric and magnetic field amplitudes of an electromagnetic wave are related to each other.
Solve: (a) Using Equation 34.29,

$$B_0 = \frac{1}{c}E_0 = \frac{100\,V/m}{3.0\times10^8\,m/s} = 3.33\times10^{-7}\,T$$

(b) From Equation 34.36, the intensity of an electromagnetic wave is

$$I = \frac{c\varepsilon_0}{2}E_0^2 = \frac{(3\times10^8\,m/s)(8.85\times10^{-12}\,C^2/N\,m^2)}{2}(100\,V/m)^2 = 13.3\,W/m^2$$

34.21. Model: A radio wave is an electromagnetic wave.
Solve: (a) The energy transported per second by the radio wave is 25 kW, or 25×10^3 J/s. This energy is carried uniformly in all directions. From Equation 34.36, the light intensity is

$$I = \frac{P}{A} = \frac{P}{4\pi r^2} = \frac{25\times10^3\,W}{4\pi(30\times10^3\,m)^2} = 2.2\times10^{-6}\,W/m^2$$

(b) Using Equation 34.36 again,

$$I = \frac{c\varepsilon_0}{2}E_0^2 \Rightarrow 2.2\times10^{-6}\,W/m^2 = \frac{(3\times10^8\,m/s)(8.85\times10^{-12}\,C^2/N\,m^2)}{2}E_0^2 \Rightarrow E_0 = 0.041\,V/m$$

34.23. Model: An object gains momentum when it absorbs electromagnetic waves.
Solve: The radiation force on an object that absorbs all the light is

$$F = \frac{P}{c} = \frac{1000\,W}{3.0\times10^8\,m/s} = 3.3\times10^{-6}\,N$$

Assess: The force is independent of the size of the beam and the wavelength.

Section 34.7 Polarization

34.25. Model: Use Malus's law for the polarized light.
Visualize:

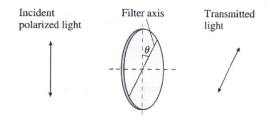

Incident
polarized light   Filter axis   Transmitted
                                                            light

θ

Solve: From Equation 34.41, the relationship between the incident and the transmitted polarized light is $I_{\text{transmitted}} = I_0 \cos^2\theta$ where θ is the angle between the electric field and the axis of the filter. Therefore,

$$0.25\, I_0 = I_0 \cos^2\theta \Rightarrow \cos\theta = 0.50 \Rightarrow \theta = 60°$$

Assess: Note that θ is the angle between the electric field and the axis of the filter.

34.29. Model: Assume the electric and magnetic fields are uniform.
Solve: The force on the proton, which is the sum of the electric and magnetic forces, is

$$\vec{F} = \vec{F}_E + \vec{F}_B = -F\cos 30°\hat{i} + F\sin 30°\hat{j} = (-2.77\hat{i} + 1.60\hat{j})\times 10^{-13}\ \text{N}$$

Since $\vec{v}$ points out of the page, the magnetic force is $\vec{F}_B = e\vec{v}\times\vec{B} = 1.60\times 10^{-13}\,\hat{j}$ N. Thus

$$\vec{F}_E = e\vec{E} = \vec{F} - \vec{F}_B = -2.77\times 10^{-13}\,\hat{i}\ \text{N} \Rightarrow \vec{E} = \vec{F}_E/e = -1.73\times 10^6\,\hat{i}\ \text{V/m}$$

That is, the electric field is $\vec{E} = (1.73\times 10^6\ \text{V/m, left})$.

34.31. Model: Use the Galilean transformation of fields. Assume that the electric and magnetic fields are uniform inside the capacitor.
Visualize: The laboratory frame is the A frame and the proton's frame is the B frame.
Solve: (a) The electric field is directed downward, and thus the electric force on the proton is downward. The magnetic field $\vec{B}$ is oriented so that the force on the proton is directed upward. Use of the right-hand rule tells us that the magnetic field is directed into the page. The magnitude of the magnetic field is obtained from setting the magnetic force equal to the electric force, yielding the equation $evB = eE$. Solving for B,

$$B = \frac{E}{v} = \frac{1.0\times 10^5\ \text{V/m}}{1.0\times 10^6\ \text{m/s}} = 0.10\ \text{T}$$

Thus $\vec{B} = (0.10\ \text{T, into page})$.
(b) In the B frame, the magnetic and electric fields are

$$\vec{B}_B = \vec{B}_A - \frac{1}{c^2}\vec{v}_{BA}\times\vec{E}_A = -0.10\hat{k}\ \text{T} - \frac{(1.0\times 10^6\,\hat{j}\ \text{m/s})\times(1.0\times 10^5\,\hat{i}\ \text{V/m})}{(3.0\times 10^8\ \text{m/s})^2} \approx -0.10\hat{k}\ \text{T}$$

$$\vec{E}_B = \vec{E}_A + \vec{v}_{BA}\times\vec{B}_A = 1.0\times 10^5\,\hat{i}\ \frac{\text{V}}{\text{m}} + (1.0\times 10^6\,\hat{j}\ \text{m/s})\times(-0.10\hat{k}\ \text{T}) = 0\ \frac{\text{V}}{\text{m}}$$

(c) There is no electric force in the proton's frame because $E_B = 0$, and there is no magnetic force because the proton is at rest in the B frame.

34.33. Model: Use the Galilean transformation of fields.
Visualize:

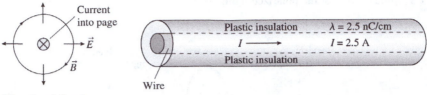

A current of 2.5 A flows to the right through the wire, and the plastic insulation has a charge of linear density $\lambda = 2.5$ n C/cm.
Solve: The magnetic field B at a distance r from the wire is

$$\vec{B} = \left(\frac{\mu_0 I}{2\pi r}, \text{ clockwise seen from left}\right)$$

On the other hand, $\vec{E}$ is radially out along $\hat{r}$, that is,

$$\vec{E} = \frac{\lambda}{2\pi\varepsilon_0 r}\hat{r}$$

As the mosquito is 1.0 cm from the center of the wire at the top of the wire,

$$B = \frac{(4\pi\times10^{-7}\text{ T m/A})(2.5\text{ A})}{2\pi(0.010\text{ m})} = 5.0\times10^{-5}\text{ T}$$

$$E = \frac{(2.5\times10^{-7}\text{ C/m})(2)(9.0\times10^9\text{ N m}^2/\text{C}^2)}{0.010\text{ m}} = 4.5\times10^5\frac{\text{V}}{\text{m}}$$

where the direction of $\vec{B}$ is out of the page and the direction of $\vec{E}$ is radially outward. In the mosquito's frame (let us call it M), we want $\vec{B}_M = \vec{0}$ T. Thus,

$$\vec{B}_M = \vec{B} - \frac{1}{c^2}\vec{v}\times\vec{E} = 0 \Rightarrow \vec{B} = \frac{1}{c^2}\vec{v}\times\vec{E}$$

Because $\vec{v}\times\vec{E}$ must be in the direction of $\vec{B}$ and $\vec{E}$ is radially outward, according to the right-hand rule $\vec{v}$ must be along the direction of the current. The magnitude of the velocity is

$$v = \frac{c^2 B}{E} = \frac{(3.0\times10^8\text{ m/s})^2(5.0\times10^{-5}\text{ T})}{4.5\times10^5\text{ V/m}} = 1.0\times10^7\text{ m/s}$$

The mosquito must fly at 1.0×10^7 m/s parallel to the current. This is highly unlikely to happen unless the mosquito is from Planet Krypton, like Superman.

34.39. Model: The displacement current through a capacitor is the same as the current in the connecting wires.
Solve:

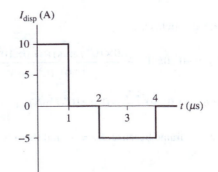

We have

$$I_{\text{disp}} = \varepsilon_0 \frac{d\Phi_e}{dt} = \varepsilon_0 \frac{d}{dt}\left(\frac{CV_c}{\varepsilon_0}\right) = C\frac{dV_c}{dt}$$

The displacement current is the slope of the V_c vs. t curve times C.

34.41. Model: The displacement current through a capacitor is the same as the current in the wires.
Solve: From Chapter 32, a discharging capacitor has circuit current

$$I = I_0 e^{-t/\tau} = \frac{(\Delta V_c)_0}{R} e^{-t/\tau}$$

Identifying $\tau = 2.0\ \mu s = RC$, we get $R = 2.0\ \Omega$. Thus

$$I_0 = \frac{(\Delta V_c)_0}{R} \Rightarrow (\Delta V_c)_0 = I_0 R = (10\ \text{A})(2.0\ \Omega) = 20\ \text{V}$$

34.43. Solve: (a) The electric field energy density is $u_E = \frac{1}{2}\varepsilon_0 E^2$ and the magnetic field energy density is $u_B = (1/2\mu_0)B^2$. In an electromagnetic wave, the fields are related by $E = cB$. Using this and the fact that $c^2 = 1/(\varepsilon_0\mu_0)$, we find

$$u_E = \frac{\varepsilon_0}{2}E^2 = \frac{\varepsilon_0}{2}(cB)^2 = \frac{c^2\varepsilon_0}{2}B^2 = \frac{\varepsilon_0}{2\varepsilon_0\mu_0}B^2 = \frac{1}{2\mu_0}B^2 = u_B$$

(b) Since the energy density is equally divided between the electric field and the magnetic field, the total energy density in an electromagnetic wave is $u_{\text{tot}} = u_E + u_B = 2u_E$. We also know that the wave intensity is $I = \frac{1}{2}c\varepsilon_0 E_0^2$. Thus

$$u_{\text{tot}} = 2u_E = \varepsilon_0 E_0^2 = \frac{2}{c}\left(\frac{1}{2}c\varepsilon_0 E_0^2\right) = \frac{2I}{c} = \frac{2(1000\ \text{W/m}^2)}{3.0\times10^8\ \text{m/s}} = 6.67\times10^{-6}\ \text{J/m}^3$$

34.45. Model: Sunlight is an electromagnetic wave.
Solve: (a) The sun's energy is transported by the electromagnetic waves in all directions. From Equation 34.36, the light intensity is

$$I = \frac{P}{A} \Rightarrow P = IA = (1360\ \text{W/m}^2)(4\pi R_{\text{sun-earth}}^2) = (1360\ \text{W/m}^2)4\pi(1.50\times10^{11}\ \text{m})^2 = 3.85\times10^{26}\ \text{W}$$

(b) The intensity of sunlight at Mars is

$$I = \frac{P}{A} = \frac{(1360\ \text{W/m}^2)(4\pi R_{\text{earth}}^2)}{4\pi R_{\text{Mars}}^2} = (1360\ \text{W/m}^2)\left(\frac{1.50\times10^{11}\ \text{m}}{2.28\times10^{11}\ \text{m}}\right)^2 = 589\ \text{W/m}^2$$

34.49. Model: Radio waves are electromagnetic waves. Assume that the transmitter unit radiates in all directions.
Solve: The transmitting unit radiates energy in all directions at the rate of 250 mJ per second. From Equation 34.36, the signal intensity at a distance of 42 m is

$$I = \frac{P}{A} = \frac{P}{4\pi r^2} = \frac{250\times10^{-3}\ \text{W}}{4\pi(42\ \text{m})^2} = 1.13\times10^{-5}\ \text{W/m}^2$$

Using Equation 34.36 again,

$$I = \frac{c\varepsilon_0}{2}E_0^2 \Rightarrow E_0 = \sqrt{\frac{2I}{c\varepsilon_0}} = \sqrt{\frac{2(1.13\times10^{-5}\ \text{W/m}^2)}{(3.0\times10^8\ \text{m/s})(8.85\times10^{-12}\ \text{C}^2/\text{N m}^2)}} = 0.092\ \frac{\text{V}}{\text{m}}$$

A few steps before 42 m, the field strength was 0.100 V/m and the door opened. The manufacturer's claims are correct.

34.53. Visualize: Use subscript 1 for the 27 MHz waves and subscript 2 for the 2.4 GHz waves. Also recall that for EM waves $c = \lambda f$.

Solve: $d \propto \lambda^{1/2}$.

$$\frac{d_2}{d_1} = \left(\frac{\lambda_2}{\lambda_1}\right)^{1/2} = \left(\frac{c/f_2}{c/f_1}\right)^{1/2} = \sqrt{\frac{f_2}{f_1}}$$

$$d_2 = d_1\sqrt{\frac{f_2}{f_1}} = (14 \text{ cm})\sqrt{\frac{2.4 \text{ GHz}}{27 \text{ MHz}}} = 132 \text{ cm} \approx 1.3 \text{ m}$$

Assess: 1.3 m is farther than most of us are thick, so the 2.4 GHz waves go through people fairly well.

34.55. Model: Assume that the black paper absorbs the light completely. Use the particle model for the paper.
Visualize:

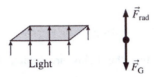

Light

For the black paper to be suspended, the radiation-pressure force must be equal to the gravitational force on the paper.
Solve: From Equation 34.38, $F_{rad} = p_{rad}A = IA/c$. Hence,

$$I = \frac{c}{A}F_{rad} = \frac{c}{A}F_G = \frac{(3.0 \times 10^8 \text{ m/s})(1.0 \times 10^{-3} \text{ kg})(9.8 \text{ m/s}^2)}{(8.5 \text{ inch} \times 11 \text{ inch})(2.54 \times 10^{-2} \text{ m/inch})^2} = 4.9 \times 10^7 \text{ W/m}^2$$

AC CIRCUITS

Exercises and Problems

Section 35.1 AC Sources and Phasors

35.3. Model: A phasor is a vector that rotates counterclockwise around the origin at angular velocity ω.
Solve: The emf is

$$\varepsilon = \varepsilon_0 \cos(\omega t) = (50 \text{ V})\cos[(2\pi \times 110 \text{ rad/s})(3.0 \times 10^{-3}\text{s})] = (50 \text{ V})\cos(2.074 \text{ rad}) = (50 \text{ V})\cos(119°)$$

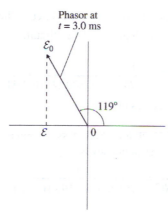

Phasor at
$t = 3.0$ ms

35.5. Visualize: Please refer to Figure 35.4 for an AC resistor circuit.
Solve: (a) For a circuit with a single resistor, the peak current is

$$I_R = \frac{\varepsilon_0}{R} = \frac{10 \text{ V}}{200 \text{ }\Omega} = 50 \text{ mA}$$

(b) The peak current is the same as in part (a) because the current is independent of frequency.

Section 35.2 Capacitor Circuits

35.7. Visualize: Figure 35.7 shows a simple one-capacitor circuit.
Solve: (a) The capacitive reactance at $\omega = 2\pi f = 2\pi(100 \text{ Hz}) = 628.3 \text{ rad/s}$ is

$$X_C = \frac{1}{\omega C} = \frac{1}{(628.3 \text{ rad/s})(0.30 \times 10^{-6} \text{ F})} = 5305 \text{ }\Omega$$

$$I_C = \frac{V_C}{X_C} = \frac{10 \text{ V}}{5.305 \times 10^3 \text{ }\Omega} = 1.88 \times 10^{-3} \text{ A} = 1.9 \text{ mA}$$

(b) The capacitive reactance at $\omega = 2\pi(100 \text{ kHz}) = 628{,}300 \text{ rad/s}$ is

$$X_C = \frac{1}{\omega C} = \frac{1}{(6.283 \times 10^5 \text{ rad/s})(0.30 \times 10^{-6} \text{ F})} = 5.305 \; \Omega$$

$$I_C = \frac{V_C}{X_C} = \frac{10 \text{ V}}{5.305 \; \Omega} = 1.9 \text{ A}$$

Assess: Using reactance is just like using resistance in Ohm's law. Because $X_C \propto \omega^{-1}$, X_C decreases with an increase in ω, as observed above.

35.9. Visualize: Figure 35.7 shows a simple one-capacitor circuit.
Solve: **(a)** From Equation 35.11,

$$I_C = \frac{V_C}{X_C} = \frac{V_C}{1/\omega C} = \omega C V_C = 2\pi f C V_C \quad \Rightarrow \quad f = \frac{50 \times 10^{-3} \text{ A}}{2\pi(5.0 \text{ V})(20 \times 10^{-9} \text{ F})} = 80 \text{ kHz}$$

(b) The AC current through a capacitor *leads* the capacitor voltage by $90°$ or $\pi/2$ rad. For a simple one-capacitor circuit $i_C = I_C \cos(\omega t + \frac{1}{2}\pi)$. For $i_C = I_C$, $(\omega t + \frac{1}{2}\pi)$ *must* be equal to $2n\pi$, where $n = 1, 2, \ldots$. This means

$$\omega t = \frac{3\pi}{2}, \frac{7\pi}{2}, \frac{11\pi}{2}, \ldots$$

At these values of ωt, $v_C = V_C \cos(\omega t) = 0 \text{ V}$. That is, i_C is maximum when $v_C = 0.0 \text{ V}$.

Section 35.3 *RC* Filter Circuits

35.13. Model: The crossover frequency occurs for a series *RC* circuit when $V_R = V_C$.
Visualize: Please refer to Figure 35.12b for the high-pass filter circuit.
Solve: From Equation 35.15,

$$\omega_c = 2\pi f_c = \frac{1}{RC} \quad \Rightarrow \quad C = \frac{1}{2\pi R f_c} = \frac{1}{2\pi(100 \; \Omega)(1000 \text{ Hz})} = 1.6 \times 10^{-6} \text{ F} = 1.6 \; \mu\text{F}$$

Assess: The output for a high-pass filter is across the resistor.

35.17. Visualize: Please refer to Figure 35.12b for a high-pass *RC* filter.
Solve: **(a)** From Equation 35.15, the crossover frequency is

$$f_c = \frac{1}{2\pi RC} = \frac{1}{2\pi(100 \; \Omega)(1.59 \times 10^{-6} \text{ F})} = 1000 \text{ Hz}$$

(b) The resistor voltage in an *RC* circuit is

$$V_R = \frac{\mathcal{E}_0 R}{\sqrt{R^2 + 1/(\omega^2 C^2)}} = \frac{\mathcal{E}_0}{\sqrt{1 + 1/(\omega^2 R^2 C^2)}}$$

where, in the last step, we factored R^2 out from the square root. Using the definition $\omega_c = 1/(RC)$, we can write the resistor voltage as

$$V_R = \frac{\mathcal{E}_0}{\sqrt{1 + 1/(\omega^2 R^2 C^2)}} = \frac{\mathcal{E}_0}{\sqrt{1 + \omega_c^2/\omega^2}} = \frac{\mathcal{E}_0}{\sqrt{1 + (f_c/f)^2}}$$

Using $\mathcal{E}_0 = 5.00 \text{ V}$, we find

f	V_R (V)
$\frac{1}{2}f_c$	2.24
f_c	3.53
$2f_c$	4.47

Assess: As expected of a high-pass filter, the voltage increases with increasing frequency.

Section 35.4 Inductor Circuits

35.19. Visualize: Figure 35.13b shows a simple one-inductor circuit.
Solve: (a) The peak current through the inductor is

$$I_L = \frac{V_L}{X_L} = \frac{V_L}{\omega L} = \frac{V_L}{2\pi fL} = \frac{10 \text{ V}}{2\pi(100 \text{ Hz})(20\times10^{-3} \text{ H})} = 0.80 \text{ A}$$

(b) At a frequency of 100 kHz instead of 100 Hz as in part (a), the reactance will increase by a factor of 1000 so the current will decrease by a factor of 1000. Thus, $I_L = 0.80$ mA.

35.20. Solve: For a simple single-inductor circuit,

$$I_L = \frac{V_L}{X_L} = \frac{V_L}{\omega L} = \frac{V_L}{2\pi fL} \implies L = \frac{V_L}{2\pi fI_L} = \frac{\sqrt{2}V_{\text{rms}}}{2\pi fI_L} = \frac{\sqrt{2}(6.0 \text{ V})}{2\pi(15\times10^3 \text{ Hz})(65\times10^{-3} \text{ A})} = 1.4\times10^{-3} \text{ H} = 1.4 \text{ mH}$$

Section 35.5 The Series *RLC* Circuit

35.23. Solve: (a) From Equation 35.30, the resonance frequency is

$$\omega_0 = \frac{1}{\sqrt{LC}}$$

When the resistance is doubled, the resonance frequency stays the same because f is independent of R. Hence, $f = 200$ kHz.
(b) From Equation 35.30,

$$f_0 = \frac{\omega_0}{2\pi} = \frac{1}{2\pi}\frac{1}{\sqrt{LC}}$$

When the capacitor value is doubled,

$$f_0' = \frac{1}{2\pi}\frac{1}{\sqrt{L(2C)}} = \frac{f_0}{\sqrt{2}} = \frac{200 \text{ kHz}}{\sqrt{2}} = 141 \text{ kHz}$$

35.25. Model: At the resonance frequency, the current in a series *RLC* circuit is a maximum. The resistor does not affect the resonance frequency.
Solve: From Equation 35.30,

$$f_0 = \frac{\omega_0}{2\pi} = \frac{1}{2\pi}\sqrt{\frac{1}{LC}} \implies C = \frac{1}{4\pi^2 f_0^2 L} = \frac{1}{4\pi^2(1000 \text{ Hz})^2(20\times10^{-3} \text{ H})} = 1.27\times10^{-6} \text{ F} \approx 1.3 \text{ } \mu\text{F}$$

35.27. Visualize: The circuit looks like that shown in Figure 35.16.
Solve: (a) The impedance of the circuit for a frequency of 3000 Hz is

$$Z = \sqrt{R^2 + (X_L - X_C)^2} = \sqrt{(50 \text{ }\Omega)^2 + \left[2\pi(3000 \text{ Hz})(3.3\times10^{-3} \text{ H}) - \frac{1}{2\pi(3000 \text{ Hz})(480\times10^{-9} \text{ F})}\right]^2}$$

$$= \sqrt{(50 \text{ }\Omega)^2 + (62.20 \text{ }\Omega - 110.52 \text{ }\Omega)^2} = 69.53 \text{ }\Omega \approx 70 \text{ }\Omega$$

The peak current is

$$I = \frac{\mathcal{E}_0}{Z} = \frac{5.0 \text{ V}}{69.53 \text{ }\Omega} = 0.072 \text{ A} = 72 \text{ mA}$$

The phase angle is

$$\phi = \tan^{-1}\left[\frac{X_L - X_C}{R}\right] = \tan^{-1}\left(\frac{-48.32 \text{ }\Omega}{50 \text{ }\Omega}\right) = -44°$$

(b) For 4000 Hz, $Z = 50$ Ω, $I = 0.10$ A, and $\phi = 0.0°$.
(c) For 5000 Hz, $Z = 62$ Ω, $I = 0.080$ A, and $\phi = 37°$

The following table summarizes the results.

	$f = 3000$ Hz	$f = 4000$ Hz	$f = 5000$ Hz
$Z\,[\Omega]$	70	50	62
$I\,[A]$	0.072	0.10	0.080
ϕ	$-44°$	$0.0°$	$37°$

Section 35.6 Power in AC Circuits

35.31. Solve: From Equation 35.45, we see that the power delivered by a source is related to the maximum power as

$$P_{\text{source}} = P_{\text{max}} \cos^2 \phi$$

If $P_{\text{source}} = 0.75\, P_{\text{max}}$, then the phase angle ϕ is

$$\cos^2 \phi = 0.75 \quad \Rightarrow \quad \phi = \cos^{-1}(\sqrt{0.75}) = 30°$$

35.35. Solve: (a) From Equation 35.14,

$$V_R = \frac{\mathcal{E}_0 R}{\sqrt{R^2 + (\omega C)^{-2}}} = \frac{\mathcal{E}_0}{2} \quad \Rightarrow \quad R^2 + \frac{1}{\omega^2 C^2} = 4R^2 \quad \Rightarrow \quad \omega = \frac{1}{\sqrt{3}\,RC}$$

(b) At this frequency,

$$V_C = IX_C = \frac{V_R}{R}\left(\frac{1}{\omega C}\right) = \frac{(\mathcal{E}_0/2)}{R}(\sqrt{3}RC)\frac{1}{C} = \frac{\sqrt{3}}{2}\mathcal{E}_0$$

35.37. Visualize: Please refer to Figure P35.37.
Solve: (a) The voltage across the capacitor is

$$V_C = IX_C = \frac{\mathcal{E}_0}{\sqrt{R^2 + X_C^2}} X_C = \frac{\mathcal{E}_0/(\omega C)}{\sqrt{R^2 + (\omega C)^{-2}}} = \frac{\mathcal{E}_0}{\sqrt{(\omega RC)^2 + 1}}$$

$$= \frac{10\text{ V}}{\sqrt{4\pi^2 f^2 (16\ \Omega)^2 (1.0\times10^{-6}\text{ F})^2 + 1}} = \frac{10\text{ V}}{\sqrt{1 + (1.0106\times10^{-8}\text{ s}^2)f^2}}$$

The values of V_C at a few frequencies are in the following table.

f (kHz)	V_C(V)
1	9.95
3	9.57
10	7.05
30	3.15
100	0.990

(b)

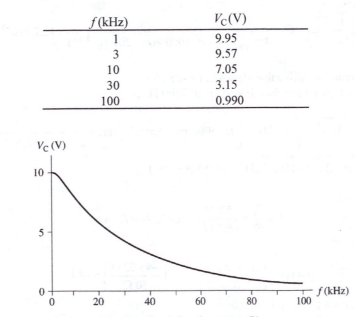

Assess: For the voltage across the capacitor, the circuit is a low-pass filter.

35.41. Visualize: Figure 35.17 defines the phase angle ϕ.
Solve: The phase angle is

$$\phi = \tan^{-1}\left(\frac{X_L - X_C}{R}\right)$$

A capacitor-only circuit has no resistor ($R = 0\,\Omega$) and no inductor ($X_L = 0\,\Omega$). Thus, the phase angle is

$$\phi = \tan^{-1}\left(\frac{0 - X_C}{0}\right) = \tan^{-1}(-\infty) = -\frac{\pi}{2}\ \text{rad}$$

That is, the current leads the voltage by $\pi/2$ rad or 90°. That is exactly the expected behavior for a capacitor circuit.

35.43. Solve: Because $Q = CV$ (and C is constant), Equation 31.31 maybe transformed to give the voltage of a discharging capacitor as

$$V_C = V_0 e^{-t/(RC)} \Rightarrow \frac{V_0}{2} = V_0 e^{-(2.5\,\text{ms})/(RC)} \quad \Rightarrow \quad \ln\left(\tfrac{1}{2}\right) = -\frac{(2.5\,\text{ms})}{RC} \quad \Rightarrow \quad RC = \frac{-(2.5 \times 10^{-3}\,\text{s})}{\ln 0.5} = 3.61 \times 10^{-3}\,\text{s}$$

The crossover frequency for a low-pass circuit is

$$f_c = \left(\frac{1}{2\pi}\right)\omega_c = \left(\frac{1}{2\pi}\right)\frac{1}{RC} = \frac{1}{2\pi(3.61 \times 10^{-3}\,\text{s})} = 44\ \text{Hz}$$

35.47. Model: While the AC current through an inductor lags the inductor voltage by 90°, the current and the voltage are in phase for a resistor.
Visualize: Series elements have the same current, so we start with a common current phasor I for the inductor and resistor,

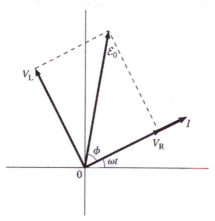

Please refer to Figure P35.47.
Solve: Because we have a series RL circuit, the current through the resistor and the inductor is the same. The voltage phasor V_R is along the same direction as the current phasor I. The voltage phasor V_L is ahead of the current phasor by 90°.

(a) From the phasors in the figure, $\mathcal{E}_0 = \sqrt{V_L^2 + V_R^2}$. Noting that $V_L = I\omega L$ and $V_R = IR$, $\mathcal{E}_0 = I\sqrt{\omega^2 L^2 + R^2}$ and

$$I = \frac{\mathcal{E}_0}{\sqrt{R^2 + \omega^2 L^2}} \qquad V_R = \frac{\mathcal{E}_0 R}{\sqrt{R^2 + \omega^2 L^2}} \qquad V_L = \frac{\mathcal{E}_0 \omega L}{\sqrt{R^2 + \omega^2 L^2}}$$

(b) As $\omega \to 0$ rad/s, $V_R \to \mathcal{E}_0 R/R = \mathcal{E}_0$ and as $\omega \to \infty$, $V_R \to 0$ V.
(c) The RL circuit will be a low-pass filter, if the output is taken from the resistor. This is because V_R is maximum when ω is low and goes to zero when ω becomes large.
(d) At the crossover frequency, $V_L = V_R$. Hence,

$$I\omega_c L = IR \quad \Rightarrow \quad \omega_c = \frac{R}{L}$$

35.49. Visualize: The circuit looks like the one in Figure 35.16.
Solve: (a) The impedance of the circuit is

$$Z = \sqrt{R^2 + (X_L - X_C)^2} = \sqrt{(25\ \Omega)^2 + \left[2\pi(60\ \text{Hz})(0.10\ \text{H}) - \frac{1}{2\pi(60\ \text{Hz})(100\times10^{-6}\ \text{F})}\right]^2}$$

$$= \sqrt{(25\ \Omega)^2 + (37.70\ \Omega - 26.53\ \Omega)^2} = 27.4\ \Omega$$

The rms voltage is

$$\mathcal{E}_{\text{rms}} = I_{\text{rms}}\, Z = (2.5\ \text{A})(27.4\ \Omega) = 69\ \text{V}$$

(b) The phase angle is

$$\phi = \tan^{-1}\left[\frac{X_L - X_C}{R}\right] = \tan^{-1}\left(\frac{11.173\ \Omega}{25\ \Omega}\right) = 24°$$

This is an inductive circuit because ϕ is positive.
(c) The average power loss is

$$P_R = I_{\text{rms}}\mathcal{E}_{\text{rms}}\cos\phi = \frac{\mathcal{E}_{\text{rms}}^2}{R}\cos\phi = \frac{(68.5\ \text{V})^2}{25\ \Omega}\cos(24°) = 0.17\ \text{kW}$$

35.53. Model: An *RLC* circuit is driven above the resonance frequency when the circuit current lags the emf.
Visualize: The circuit looks like the one in Figure 35.16.
Solve: The phase angle is $+30°$ since the circuit is above resonance, and $X_L > X_C$. Thus

$$\tan 30° = \frac{X_L - X_C}{R} \quad \Rightarrow \quad X_L - X_C = R\tan 30°$$

The impedance of the circuit is

$$Z = \sqrt{R^2 + (X_L - X_C)^2} = \sqrt{R^2 + (R\tan 30°)^2} = R\sqrt{1 + (\tan 30°)^2}$$

The peak current is

$$I = \frac{\mathcal{E}_0}{Z} = \frac{10\ \text{V}}{(50\ \Omega)\sqrt{1 + (\tan 30°)^2}} = 0.17\ \text{A}$$

Assess: Remember to put your calculator back into degrees mode when calculating $\tan 30°$.

35.55. Visualize: The circuit looks like the one in Figure 35.16.
Solve: (a) The instantaneous current is $i = I\cos(\omega t - \phi)$. The phase angle is

$$\phi = \tan^{-1}\left[\frac{X_L - X_C}{R}\right] = \tan^{-1}\left(\frac{-48.32\ \Omega}{50\ \Omega}\right) = -44°$$

Since $i = I\cos(\omega t - \phi)$, $i = I$ implies that $\omega t - \phi = 0$ rad. That is, $\omega t = \phi = -44°$. Thus,

$$\mathcal{E} = \mathcal{E}_0 \cos\omega t = (5.0\ \text{V})\cos(-44°) = 3.6\ \text{V}$$

(b) $i = 0$ A and is decreasing implies that $\omega t - \phi = \frac{1}{2}\pi$ rad. That is, $\omega t = \frac{1}{2}\pi + \phi = 90° - 44° = 46°$. Thus, $\mathcal{E} = (5.0\ \text{V})\cos(46°) = 3.5\ \text{V}$.
(c) $i = -I$ imples $\omega t - \phi = \pi$ rad. That is, $\omega t = \pi + \phi = 180° - 44° = 136°$. Thus, $\mathcal{E} = (5.0\ \text{V})\cos 136° = -3.6\ \text{V}$.

35.57. Visualize: Consider the phasor diagram shown in the figure below.

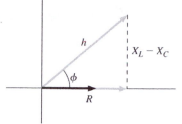

Solve: From Equation 35.27, we know that phase angle ϕ is given by

$$\phi = \tan^{-1}\left(\frac{X_L - X_C}{R}\right)$$

as shown in the figure above. By the Pythagorean theorem, the hypotenuse h is

$$h = \sqrt{R^2 + (X_L - X_C)^2} = Z$$

The definition of $\cos\phi$ then gives

$$\cos\phi = \frac{R}{h} = \frac{R}{Z}$$

Assess: Note that $\cos\phi$ is dimensionless, as expected.

35.61. Model: The filament in a light bulb acts as a resistor.
Visualize: Please refer to Figure P35.61.
Solve: A bulb labeled 40 W is designed to dissipate an average of 40 W of power at a voltage of $V_{rms} = 120$ V. From Equation 35.39, the resistance of the 40 W light bulb is

$$R_{40} = \frac{V_{rms}^2}{P_{40}} = \frac{(120 \text{ V})^2}{40 \text{ W}} = 360 \text{ }\Omega$$

Likewise, $R_{60} = 240 \text{ }\Omega$ and $R_{100} = 144 \text{ }\Omega$. The rms current through the 40 W and 60 W bulbs, which are in series, is

$$I_{60} = I_{40} = \frac{\mathcal{E}_0}{R_{40} + R_{60}} = \frac{\mathcal{E}_0}{600 \text{ }\Omega} = \frac{120 \text{ V}}{600 \text{ }\Omega} = 0.20 \text{ A}$$

The rms voltage across the light bulbs in series are

$$V_{40} = I_{40}R_{40} = (0.20 \text{ A})(360 \text{ W}) = 72 \text{ V} \qquad V_{60} = I_{60}R_{60} = (0.20 \text{ A})(240 \text{ W}) = 48 \text{ V}$$

The voltage across the 100 W light bulb is $V_{100} = 120$ V. The powers dissipated in the light bulbs in series are

$$P_{40} = V_{40}I_{40} = (72 \text{ V})(0.20 \text{ A}) = 14 \text{ W} \qquad P_{60} = V_{60}I_{60} = (48 \text{ V})(0.20 \text{ A}) = 9.6 \text{ W}$$

The power dissipated in the 100 W bulb is

$$P_{100} = \frac{V_{100}^2}{R_{100}} = \frac{(120 \text{ V})^2}{144 \text{ }\Omega} = 100 \text{ W}$$

RELATIVITY

Exercises and Problems

Section 36.2 Galilean Relativity

36.1. Model: S and S′ are inertial frames that overlap at $t = 0$. Frame S′ moves with a speed $v = 5.0$ m/s along the x-direction relative to frame S.
Visualize:

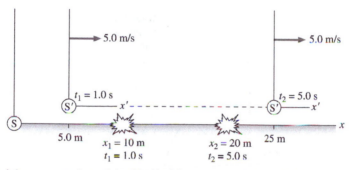

The figure shows a pictorial representation of the S and S′ frames at $t = 1.0$ s and 5.0 s.
Solve: From the figure, the observer in S′ finds the position of the first explosion at $x_1' = -5.0$ m at $t = 1.0$ s. The position of the second explosion is $x_2' = -5.0$ m at $t = 5.0$ s. We can get the same answers using the Galilean transformations of position:

$$x_1' = x_1 - vt = 10 \text{ m} - (5.0 \text{ m/s})(1.0 \text{ s}) = 5.0 \text{ m at } 1.0 \text{ s}$$
$$x_2' = x_2 - vt = 20 \text{ m} - (5.0 \text{ m/s})(5.0 \text{ s}) = -5.0 \text{ m at } 5.0 \text{ s}$$

36.5. Model: The boy on a bicycle is frame S′ and the ground is frame S. S′ moves relative to S with a speed $v = 5.0$ m/s. The frames S and S′ overlap at $t = 0$.
Visualize:

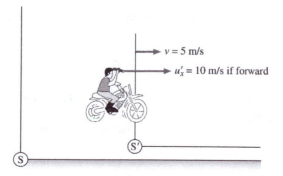

The figure shows a pictorial representation of the two frames.

Solve: **(a)** When the newspaper is thrown forward, $u'_x = 8.0$ m/s. The Galilean transformation of velocity is

$$u_x = u'_x + v = 8.0 \text{ m/s} + 5.0 \text{ m/s} = 13 \text{ m/s}$$

(b) When the newspaper is thrown backward, $u_y = u'_y = 8.0$ m/s. In this case

$$u_x = u'_x + v = -8.0 \text{ m/s} + 5.0 \text{ m/s} = -3.0 \text{ m/s}$$

Thus the speed is 3.0 m/s.

(c) When the newspaper is thrown to the side, $u_y = u'_y = 8.0$ m/s. Also,

$$u = \sqrt{u_x^2 + u_y^2} = \sqrt{(5.0 \text{ m/s})^2 + (8.0 \text{ m/s})^2} = 9.4 \text{ m/s}$$

Thus the newspaper's speed is

$$u = \sqrt{u_x^2 + u_y^2} = \sqrt{(5.0 \text{ m/s})^2 + (8.0 \text{ m/s})^2} = 9.4 \text{ m/s}$$

Section 36.4 Events and Measurements

Section 36.5 The Relativity of Simultaneity

36.9. Model: The clocks are in the same reference frame.
Visualize:

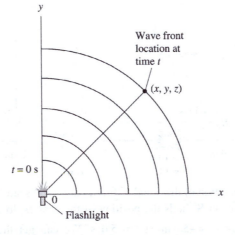

Solve: The speed of light is $c = 300$ m/μs $= 0.30$ m/ns. The distance from the origin to the point $(x, y, z) =$ $(30 \text{ m}, 40 \text{ m}, 0 \text{ m})$ is $\sqrt{(30 \text{ m})^2 + (40 \text{ m})^2} = 50$ m. So, the time taken by the light to travel 50 m is

$$\frac{50 \text{ m}}{0.30 \text{ m/ns}} = 167 \text{ ns}$$

The clock should be preset to 167 ns.

36.11. Model: Bianca and firecrackers 1 and 2 are in the same reference frame. Light from both firecrackers travels toward Bianca at 300 m/μs.
Visualize:

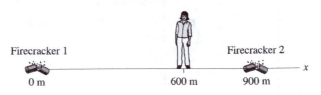

Solve: The flash from firecracker 1 takes 2.0 μs to reach Bianca (600 m ÷ 300 m/μs). The firecracker exploded at $t_1 = 1.0$ μs because it reached Bianca's eye at 3.0 μs. The flash from the firecracker 2 takes 1.0 μs to reach Bianca. Since firecrackers 1 and 2 exploded simultaneously, the explosion occurs at $t_2 = 1.0$ μs. So, the light from firecracker 2 reaches Bianca's eye at 2.0 μs. Although the events are simultaneous, Bianca *sees* them occurring at different times.

36.15. Model: Your personal rocket craft is an inertial frame moving at 0.9c relative to stars A and B.
Solve: In your frame, star A is moving away from you and star B is moving toward you. When you are exactly halfway between them, both the stars explode simultaneously. The flashes from the two stars travel toward you with speed c. Because (i) you are at rest in your frame, (ii) the explosions are equally distant, and (iii) the light speed is c, independent of the fact that the stars are moving in your frame, the light will arrive simultaneously.

Section 36.6 Time Dilation

36.17. Model: Let the moving clock be in frame S' and an identical at-rest clock be in frame S.
Solve: The ticks being measured are those of the moving clock. The interval between 2 ticks is measured by the same clock in S'—namely, the clock that is ticking—so this is the proper time: $\Delta t' = \Delta \tau$. The rest clock measures a longer interval Δt between two ticks of the moving clock. These are related by

$$\Delta \tau = \sqrt{1 - \beta^2} \, \Delta t$$

The moving clock ticks at half the rate of the rest clock when $\Delta \tau = \frac{1}{2} \Delta t$. Thus

$$\sqrt{1 - \beta^2} = \sqrt{1 - v^2/c^2} = 1/2 \Rightarrow v = c\sqrt{1 - (1/2)^2} = 0.866c$$

36.19. Model: Let S be the earth's reference frame and S' the rocket's reference frame.
Solve: (a) The astronauts measure proper time i $\Delta t' = \Delta \tau$. Thus

$$\Delta t = \frac{\Delta \tau}{\sqrt{1 - (v/c)^2}} \Rightarrow 120 \text{ y} = \frac{10 \text{ y}}{\sqrt{1 - (v/c)^2}} \Rightarrow v = 0.9965c$$

(b) In frame S, the distance of the distant star is
$$\Delta x = v\Delta t = (0.9965c)(60 \text{ y}) = (0.9965 \text{ ly/y})(60 \text{ y}) = 59.8 \text{ ly}$$

36.21. Model: The ground's frame is S and the moving clock's frame is S'.
Visualize: $\Delta t = 1.0 \text{ d} + 1.0 \text{ ns} = 86400.000000001$ s and $\Delta t' = 1.0 \text{ d} = 86,400$ s.
Solve: We want to solve for v in

$$\Delta t = \gamma \Delta t'$$

Using the binomial approximation for γ.

$$\frac{\Delta t}{\Delta t'} \approx 1 + \frac{1}{2}\frac{v^2}{c^2}$$

$$\frac{\Delta t}{\Delta t'} - 1 \approx \frac{1}{2}\frac{v^2}{c^2} \Rightarrow 2c^2\left(\frac{\Delta t}{\Delta t'} - 1\right) = v^2$$

$$v = c\sqrt{2\left(\frac{\Delta t}{\Delta t'} - 1\right)} = c\sqrt{2\left(\frac{1.0 \text{ d} + 1.0 \text{ ns}}{1.0\text{d}} - 1\right)} = c\sqrt{2\left(\frac{1.0 \text{ ns}}{1.0\text{d}}\right)}$$

$$v = (3.0 \times 10^8 \text{ m/s})\sqrt{2\left(\frac{1.0 \times 10^{-9} \text{ s}}{86,400 \text{ s}}\right)} = 46 \text{ m/s}$$

Assess: This speed is about 100 mph, which is certainly doable. The calculation would be difficult without the binomial approximation due to limited calculator precision; fortunately, the approximation is excellent in this case.

Section 36.7 Length Contraction

36.27. Model: S is the ground's reference frame and S′ is the meter stick's reference frame. In the S′ frame, which moves with a velocity v relative to S, the length of the meter stick is the proper length because the meter stick is at rest in this frame. So $L′ = \ell$.

Solve: An experimenter on the ground measures the length to be contracted to

$$L = \sqrt{1 - \beta^2}\,\ell \approx \left(1 - \frac{1}{2}\beta^2\right)\ell \Rightarrow \text{shrinking} = \ell - L = \frac{1}{2}\beta^2 \ell$$

Thus the speed is

$$\beta = \sqrt{\frac{2(\ell - L)}{\ell}} = \sqrt{\frac{2(50 \times 10^{-6}\ \text{m})}{1.00\ \text{m}}} = 0.01 \Rightarrow v = \beta c = 3.0 \times 10^6\ \text{m/s}$$

Section 36.8 The Lorentz Transformations

36.29. Solve: We have $\gamma = [1 - (v/c)^2]^{-\frac{1}{2}} = [1 - (0.60)^2]^{-\frac{1}{2}} = 1.25$. In the earth's reference frame, the Lorentz transformations yield

$$x = \gamma(x′ + vt′) = 1.25[3.0 \times 10^{10}\ \text{m} + (0.60)(3.0 \times 10^8\ \text{m/s})(200)] = 8.25 \times 10^{10}\ \text{m} \approx 8.3 \times 10^{10}\ \text{m}$$

$$t = \gamma\left(t′ + \frac{vx′}{c^2}\right) = 1.25\left[200 + \frac{(0.60)(3.0 \times 10^8\ \text{m/s})(3.0 \times 10^{10}\ \text{m})}{(3.0 \times 10^8\ \text{m/s})^2}\right] = 325\ \text{s} \approx 330\ \text{s}$$

36.31. Model: The rocket and the earth are inertial frames. Let the earth be frame S and the rocket be frame S′. S′ moves with $v = 0.8c$ relative to S. The bullet's velocity in reference frame S′ is $u′ = -0.9c$.

Solve: Using the Lorentz velocity transformation equation,

$$u = \frac{u′ + v}{1 + u′v/c^2} = \frac{-0.9c + 0.8c}{1 + (-0.9c)(0.8c)/c^2} = -0.36c$$

The bullet's speed is $0.36c$. Note that the velocity transformations use *velocity*, which can be negative, and not speed.

36.33. Model: The earth and the other galaxy are inertial reference frames. Let the earth be frame S and the other galaxy be frame S′. S′ moves with $v = +0.2c$. The quasar's speed in frame S is $u = +0.8c$.

Solve: Using the Lorentz velocity transformation equation,

$$u′ = \frac{u - v}{1 - uv/c^2} = \frac{0.8c - 0.2c}{1 - (0.8c)(0.2c)/c^2} = 0.71c$$

Assess: In Newtonian mechanics, the Galilean transformation of velocity would give $u′ = 0.6c$.

Section 36.9 Relativistic Momentum

36.37. Solve: We have

$$p = \frac{mu}{\sqrt{1 - u^2/c^2}} = mc \Rightarrow \sqrt{1 - u^2/c^2} = \frac{u}{c} \Rightarrow 1 = \frac{2u^2}{c^2} \Rightarrow u = \frac{c}{\sqrt{2}} = 0.707c$$

Assess: The particle's momentum being equal to mc does not mean that the particle is moving with the speed of light. We must use the relativistic formula for the momentum as the particle speeds become high.

Section 36.10 Relativistic Energy

36.41. Solve: Equation 36.42 is $E = \gamma_{\text{p}} mc^2 = E_0 + K$. For $K = 2E_0$,

$$\gamma_{\text{p}} mc^2 = E_0 + 2E_0 = 3mc^2 \Rightarrow \gamma_{\text{p}} = \frac{1}{\sqrt{1 - u^2/c^2}} = 3 \Rightarrow 1 - \frac{u^2}{c^2} = \frac{1}{9} \Rightarrow u = \frac{\sqrt{8}}{3}c = 0.943c$$

36.43. Model: Let S be the laboratory frame and S′ be the reference frame of the 100 g ball. S′ moves to the right with a speed of $v = 2.0$ m/s relative to frame S. The 50 g ball's speed in frame S is $u_{50} = 4.0$ m/s. Because these speeds are much smaller than the speed of light, we can use the Galilean transformations of velocity.
Visualize:

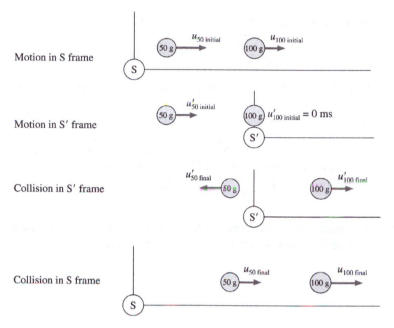

Solve: Transform the collision from frame S into frame S′, where $u'_{100 \text{ initial}} = 0$ m/s. Using the Galilean velocity transformation,

$$u'_{50 \text{ initial}} = u_{50 \text{ initial}} - v = 4.0 \text{ m/s} - 2.0 \text{ m/s} = 2.0 \text{ m/s}$$

Using Equation 10.42,

$$u'_{50 \text{ final}} = \frac{50 \text{ g} - 100 \text{ g}}{50 \text{ g} + 100 \text{ g}} u'_{50 \text{ initial}} = -\left(\frac{1}{3}\right)(2.0 \text{ m/s}) = -\left(\frac{2.0}{3}\right) \text{ m/s}$$

$$u'_{100 \text{ final}} = \frac{2(50 \text{ g})}{50 \text{ g} + 100 \text{ g}} u'_{50 \text{ initial}} = \left(\frac{2}{3}\right)(2.0 \text{ m/s}) = +\left(\frac{4.0}{3}\right) \text{ m/s}$$

Using the Galilean transformations of velocity again to go back to the S frame,

$$u_{50 \text{ final}} = u'_{50 \text{ final}} + v = -\left(\frac{2.0}{3}\right) \text{ m/s} + 2.0 \text{ m/s} = +1.33 \text{ m/s}$$

$$u_{100 \text{ final}} = u'_{100 \text{ final}} + v = \left(\frac{4.0}{3}\right) \text{ m/s} + 2.0 \text{ m/s} = +3.33 \text{ m/s}$$

Because of plus signs with $u_{50 \text{ final}}$ and $u_{100 \text{ final}}$, both masses are moving to the right.

36.47. Model: Let S be the galaxy's frame and S′ the alien spacecraft's frame. The spacetime interval s between the two events is invariant in all frames.
Solve: (a) The light from Alpha's explosion will travel 10 ly in 10 years. Since neither light nor any other signal from Alpha can travel 100 ly in 10 years to reach Beta, the explosion of Alpha could not cause the explosion of Beta.
(b) Because the spacetime interval s between the two events is invariant,

$$s^2 = c^2(\Delta t)^2 - (\Delta x)^2 = c^2(\Delta t')^2 - (\Delta x')^2 \Rightarrow \left(\frac{1 \text{ ly}}{y}\right)^2 (10 \text{ y})^2 - (100 \text{ ly})^2 = \left(\frac{1 \text{ ly}}{y}\right)^2 (\Delta t')^2 - (120 \text{ ly})^2$$

$$\Rightarrow (10 \text{ y})^2 - (100 \text{ y})^2 = (\Delta t')^2 - (120 \text{ y})^2 \Rightarrow \Delta t' = 67.1 \text{ years}$$

36.49. Model: The earth is frame S and the starship is frame S′. S′ moves relative to S with a speed v.
Solve: (a) The speed of the starship is

$$v = \frac{20 \text{ ly}}{25 \text{ y}} = \frac{(20 \text{ y})c}{25 \text{ y}} = 0.80c$$

(b) The astronauts measure the proper time while they are traveling. This is

$$\Delta \tau = \sqrt{1 - \frac{v^2}{c^2}} \Delta t = \sqrt{1 - (0.8)^2} \, (25 \text{ y}) = 15 \text{ y}$$

Because the explorers stay on the planet for one year, the time elapsed on their chronometer is 16 years.

36.51. Model: S′ is the electron's frame and S is the ground's frame. S′ moves relative to S with a speed $v = 0.99999997c$.
Solve: For an experimenter in the S frame, the length of the accelerator tube is 3.2 km. This is the proper length $\ell = L$ because it is at rest and is always there for measurements. The electron measures the tube to be length contracted to

$$L' = \sqrt{1 - \beta^2} \, \ell = \sqrt{1 - (0.99999997)^2} \, (3200 \text{ m}) = 0.78 \text{ m}$$

36.55. Model: Let S be the earth's reference frame and S′ be the reference frame of one rocket. S′ moves relative to S with $v = -0.75c$. The speed of the second rocket in the frame S is $u = 0.75c$.
Visualize:

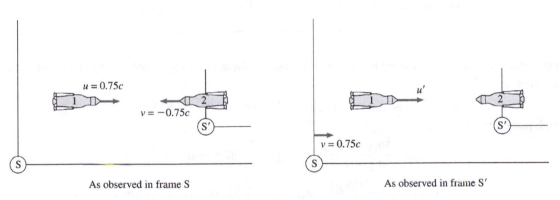

As observed in frame S   As observed in frame S′

Solve: Using the Lorentz velocity transformation equation,

$$u' = \frac{u - v}{1 - uv/c^2} = \frac{0.75c - (-0.75c)}{1 - (0.75c)(-0.75c)/c^2} = 0.96c$$

Assess: In Newtonian mechanics, the Galilean transformation of velocity will give $u' = 0.75c - (-0.75c) = 1.50c$. This is not permissible according to the theory of relativity.

36.57. Model: Use the relativistic expression for kinetic energy in Equation 36.44: $K = (\gamma_p - 1)E_0$. The electric potential energy of the electron is transformed into its kinetic energy.
Visualize: Use the conservation of energy equation $U_f - U_i + k_f - K_i = 0$ J.
Solve:

$$(-e)(V_f - V_i) + K_f - 0 \text{ J} = 0 \text{ J} \quad \Rightarrow \quad K_f = e\Delta V$$

$$\Delta V = \frac{K_f}{e} = \frac{(\gamma_p - 1)E_0}{e} = \frac{(\gamma_p - 1)mc^2}{e} = \frac{\left(\frac{1}{\sqrt{1 - 0.99^2}} - 1\right)(9.1 \times 10^{-31} \text{ kg})(3.0 \times 10^8 \text{ m/s})^2}{1.6 \times 10^{-19} \text{ C}} = 3.1 \times 10^6 \text{ V}$$

Assess: Three million volts is easily obtainable.

36.61. Model: The principle of relativity demands that all laws of physics be the same in all inertial frames.
Solve: (a) If you are in the S frame, you see the blue paint nozzle approaching at high speed. If the perpendicular lengths contract, then the blue nozzle will be less than 1 meter up from the x-axis and the blue nozzle will paint a line of blue under the red nozzle. If you are in the S′ frame, you see the red nozzle approaching at a high speed. Then, the red bar will shorten and you will see a red line under the blue nozzle.
(b) Physically you can't have both of these happen. So, we conclude that lengths perpendicular to the motion are not affected.

36.65. Model: Use Equations 36.35 and 36.42 for the momentum and total energy. Also, the quantity $E_0 = mc^2$ is an invariant in *all* inertial reference frames.
Solve: (a) The momentum and energy are

$$p = \gamma mu = \frac{1}{\sqrt{1-u^2/c^2}} mu = \frac{1}{\sqrt{1-(0.99)^2}}(1.67\times10^{-27} \text{ kg})[(0.99)(3\times10^8 \text{ m/s})] = 3.5\times10^{-18} \text{ kg m/s}$$

$$E = \gamma mc^2 = \frac{1}{\sqrt{1-u^2/c^2}} mc^2 = \frac{1}{\sqrt{1-(0.99)^2}}(1.67\times10^{-27} \text{ kg})(3.0\times10^8 \text{ m/s})^2 = 1.1\times10^{-9} \text{ J}$$

(b) From Equation 36.45, $E'^2 - p'^2 c^2 = E_0^2 = m^2 c^4$. Thus,

$$p'^2 = \frac{E'^2}{c^2} - m^2 c^2 = \frac{(5.0\times10^{-10} \text{ J})^2}{(3.0\times10^8 \text{ m/s})^2} - (1.67\times10^{-27} \text{ kg})^2 (3.0\times10^8 \text{ m/s})^2$$

$$= (2.778\times10^{-36} - 0.2510\times10^{-36}) \text{ kg}^2 \text{ m}^2/\text{s}^2 \Rightarrow p' = 1.6\times10^{-18} \text{ kg m/s}$$

36.67. Model: Mass and energy are equivalent and given by Equation 36.43.
Solve: (a) The power plant running at full capacity for 80% of the year runs for

$$(0.80)(365\times24\times3600)\text{s} = 2.52\times10^7 \text{ s}$$

The amount of thermal energy generated per year is

$$3\times(1000\times10^6 \text{ J/s})\times(2.52\times10^7 \text{ s}) = 7.56\times10^{16} \text{ J} \approx 7.6\times10^{16} \text{ J}$$

(b) Since $E_0 = mc^2$, the mass of uranium transformed into thermal energy is

$$m = \frac{E_0}{c^2} = \frac{7.56\times10^{16} \text{ J}}{(3.0\times10^8 \text{ m/s})^2} = 0.84 \text{ kg}$$

36.71. Model: Particles can be created from energy, and particles can return to energy. When a particle and its antiparticle meet, they annihilate each other and create two gamma ray photons.
Solve: The energy of the electron is

$$E_{\text{electron}} = \gamma_p m_e c^2 = \frac{1}{\sqrt{1-(0.9)^2}}(9.11\times10^{-31} \text{ kg})(3.0\times10^8 \text{ m/s})^2 = 1.88\times10^{-13} \text{ J}$$

The energy of the positron is the same, so the total energy is $E_{\text{total}} = E_{\text{electron}} + E_{\text{positron}} = 3.76\times10^{-13}$ J. The energy is converted to two equal-energy photons. Thus, $E_{\text{total}} = 2hf = 2hc/\lambda$. The wavelength is

$$\lambda = \frac{2hc}{E_{\text{total}}} = \frac{2(6.62\times10^{-34} \text{ J s})}{3.76\times10^{-13} \text{ J}} = 1.06\times10^{-12} \text{ m} \approx 1 \text{ pm}$$

Assess: This wavelength is typical of γ-ray photons.

THE FOUNDATIONS OF MODERN PHYSICS

Exercises and Problems

Section 37.2 The Emission and Absorption of Light

37.1. Model: Use Equation 37.4, which is the Balmer formula.
Visualize: Please refer to Figure 37.7.
Solve: (a) The wavelengths in the hydrogen emission spectrum are 656.6 nm, 486.3 nm, 434.2 nm, and 410.3 nm. The formula for the Balmer series can be written

$$\frac{1}{m^2} - \frac{1}{n^2} = \frac{91.18 \text{ nm}}{\lambda}$$

where $m = 1, 2, 3, \ldots$ and $n = m + 1, m + 2, \ldots$. For the first wavelength,

$$\frac{1}{m^2} - \frac{1}{n^2} = \frac{91.18 \text{ nm}}{656.5 \text{ nm}} = 0.1389 \quad \Rightarrow \quad \frac{n^2 - m^2}{n^2 m^2} = 0.1389$$

This equation is satisfied when $m = 2$ and $n = 3$. For the second wavelength (486.3 nm) the equation is satisfied for $m = 2$ and $n = 4$. Likewise, for the next two wavelengths $m = 2$ and $n = 5$ and 6.
(b) The fifth line in the spectrum will correspond to $m = 2$ and $n = 7$. Its wavelength is

$$\lambda = \frac{91.18 \text{ nm}}{\left(\frac{1}{2}\right)^2 - \left(\frac{1}{7}\right)^2} = \left(\frac{196}{45}\right)(91.18 \text{ nm}) = 397.1 \text{ nm}$$

37.5. Model: Assume the blackbodies obey Wein's law in Equation 37.2: $\lambda_{\text{peak}} = (2.90 \times 10^6 \text{ nm K})/T$.
Visualize: We want to solve for T in the equation above. We are given $\lambda_{\text{peak}} = 300$ nm, 3000 nm.
Solve:

$$T = \frac{2.90 \times 10^6 \text{ nm} \cdot \text{K}}{\lambda_{\text{peak}}}$$

(a)

$$T = \frac{2.90 \times 10^6 \text{ nm} \cdot \text{K}}{300 \text{ nm}} = 9667 \text{ K} = 9.39 \times 10^3 \, ^\circ\text{C}$$

(b)

$$T = \frac{2.90 \times 10^6 \text{ nm} \cdot \text{K}}{3000 \text{ nm}} = 966.7 \text{ K} = 694 ^\circ\text{C}$$

Section 37.3 Cathode Rays and X Rays

Section 37.4 The Discovery of the Electron

37.9. Model: Assume the fields between the electrodes are uniform and that they are zero outside the electrodes.
Visualize: Please refer to Figures 37.12 and 37.13.
Solve: (a) The speed with which a particle can pass without deflection is

$$v = \frac{E}{B} = \frac{\Delta V/d}{B} = \frac{(600 \text{ V})/(5.0 \times 10^{-3} \text{ m})}{2.0 \times 10^{-3} \text{ T}} = 6.0 \times 10^7 \text{ m/s}$$

(b) The radius of cyclotron motion in a magnetic field is

$$r = \left(\frac{m}{e}\right)\frac{v}{B} = \left(\frac{9.11 \times 10^{-31} \text{ kg}}{1.60 \times 10^{-19} \text{ C}}\right)\left(\frac{6.0 \times 10^7 \text{ m}}{2.0 \times 10^{-3} \text{ T}}\right) = 0.17 \text{ m} = 17 \text{ cm}$$

Section 37.5 The Fundamental Unit of Charge

37.11. Model: Assume the electric field ($E = \Delta V/d$) between the plates is uniform.
Visualize: Please refer to Figure 37.14.
Solve: (a) The mass of the droplet is

$$m_{\text{drop}} = \rho V = \rho\left(\frac{4\pi}{3}R^3\right) = (885 \text{ kg/m}^3)\frac{4\pi}{3}(0.4 \times 10^{-6} \text{ m})^3 = 2.37 \times 10^{-16} \text{ kg} \approx 2.4 \times 10^{-16} \text{ kg}$$

(b) In order for the upward electric force to balance the gravitational force, the charge on the droplet must be

$$Eq_{\text{drop}} = F_G = m_{\text{drop}}g$$

$$q_{\text{drop}} = \frac{m_{\text{drop}}g}{E} = \frac{(2.37 \times 10^{-16} \text{ kg})(9.8 \text{ m/s}^2)}{20 \text{ V}/11 \times 10^{-3} \text{ m}} = 1.28 \times 10^{-18} \text{ C} \approx 1.3 \times 10^{-18} \text{ C}$$

(c) Because the electric force is directed toward the electrode at the higher potential (or more positive plate), the charge on the droplet is negative. The number of surplus electrons is

$$N = \frac{q_{\text{droplet}}}{e} = \frac{1.28 \times 10^{-18} \text{ C}}{1.60 \times 10^{-19} \text{ C}} = 8$$

37.13. Model: The charge on an object is an integral multiple of a certain minimum charge value.
Solve: From smallest charge to the largest charge, the measured charges on the drops are 2.66×10^{-19} C, 3.99×10^{-19} C, 9.31×10^{-19} C, and 10.64×10^{-19} C. The differences between these values are 1.33×10^{-19} C, 2.66×10^{-19} C, 2.66×10^{-19} C, and 1.33×10^{-19} C. These differences are a multiple of 1.33×10^{-19} C. Thus, the largest value of the fundamental unit of charge that is consistent with the charge measurements is 1.33×10^{-19} C.

Section 37.6 The Discovery of the Nucleus

Section 37.7 Into the Nucleus

37.15. Model: The electron volt is a unit of energy and is defined as the kinetic energy gained by an electron or proton if it accelerates through a potential difference of 1 volt.
Solve: (a) Converting electron volts to joules,

$$7.0 \text{ MeV} = 7.0 \times 10^6 \text{ eV} \times \frac{1.6 \times 10^{-19} \text{ J}}{1.0 \text{ eV}} = 1.12 \times 10^{-12} \text{ J}$$

Using the definition of kinetic energy $K = \frac{1}{2}mv^2$, the speed of the neutron is

$$v = \sqrt{\frac{2K}{m}} = \sqrt{\frac{2(1.12 \times 10^{-12} \text{ J})}{1.67 \times 10^{-27} \text{ kg}}} = 3.7 \times 10^7 \text{ m/s}$$

(b) Likewise, the speed of the helium atom is

$$v = \sqrt{\frac{2(15 \text{ MeV})}{4(1.67 \times 10^{-27} \text{ kg})}} = \sqrt{\frac{2(2.40 \times 10^{-12} \text{ J})}{4(1.67 \times 10^{-27} \text{ kg})}} = 2.7 \times 10^7 \text{ m/s}$$

(c) The mass of the particle is

$$m = \frac{2K}{v^2} = \frac{2(1.14 \text{ keV})}{(2.0 \times 10^7 \text{ m/s})^2} = \frac{2(1.82 \times 10^{-16} \text{ J})}{(2.0 \times 10^7 \text{ m/s})^2} = 9.1 \times 10^{-31} \text{ kg}$$

The particle is an electron.

37.19. Model: For a neutral atom, the number of electrons is the same as the number of protons.
Solve: (a) Since $Z = 3$, ^{6}Li has 3 electrons, 3 protons, and $6 - 3 = 3$ neutrons.

(b) Since $Z = 8$, the ^{16}O$^+$ ion has 7 electrons, 8 protons, and $16 - 8 = 8$ neutrons.

(c) Since $Z = 7$, the doubly charged ^{13}N^{++} ion has 5 electrons, 7 protons, and $13 - 7 = 6$ neutrons.

37.23. Model: For a neutral atom, the number of electrons is the same as the number of protons, which is the atomic number Z. An atom's mass number is $A = Z + N$, where N is the number of neutrons.
Solve: (a) For a ^{197}Au atom, $Z = 79$, so $N = 197 - 79 = 118$. A neutral ^{197}Au atom contains 79 electrons, 79 protons, and 118 neutrons.
(b) Assuming that the neutron rest mass is the same as the proton rest mass, the density of the gold nucleus is

$$\rho_{\text{nucleus}} = \frac{197m_{\text{proton}}}{\frac{4\pi}{3}\left(\frac{14.0}{2} \times 10^{-15} \text{ m}\right)^3} = \frac{197(1.67 \times 10^{-27} \text{ kg})}{\frac{4\pi}{3}\left(\frac{14.0}{2} \times 10^{-15} \text{ m}\right)^3} = 2.29 \times 10^{17} \text{ kg/m}^3$$

(c) The nuclear density in part (b) is 2.01×10^{13} times the density of lead.
Assess: The mass of the matter is primarily in the nuclei and the volume of the matter is essentially due to the electrons around the nuclei.

37.27. Solve: The rest energy of an electron is

$$E_0 = mc^2 = (9.11 \times 10^{-31} \text{ kg})(3.00 \times 10^8 \text{ m/s})^2 = (8.199 \times 10^{-14} \text{ J})\left(\frac{1 \text{ eV}}{1.6 \times 10^{-19} \text{ J}}\right) = 0.512 \text{ MeV}$$

The rest energy of a proton is

$$E_0 = mc^2 = (1.67 \times 10^{-27} \text{ kg})(3.00 \times 10^8 \text{ m/s})^2 = 1.503 \times 10^{-10} \text{ J} \times \frac{1 \text{ eV}}{1.6 \times 10^{-19} \text{ J}} = 939 \text{ MeV}$$

37.29. Model: Mass and energy are equivalent.
Solve: The energy released as kinetic energy is

$$K = \Delta E = (\Delta m)c^2 = (0.185)(1.66 \times 10^{-27} \text{ kg})(3.00 \times 10^8 \text{ m/s})^2 = (2.76 \times 10^{-11} \text{ J})\left(\frac{1 \text{ eV}}{1.6 \times 10^{-19} \text{ J}}\right) = 173 \text{ MeV}$$

37.31. Model: Assume the fields between the electrodes are uniform and that they are zero outside the electrodes.
Solve: In a crossed-field experiment, the deflection is zero when the magnetic and electric forces exactly balance each other. From Equation 37.8,

$$B = \frac{E}{v} = \frac{\Delta V/d}{v} = \frac{(500 \text{ V})/(1.0 \times 10^{-2} \text{ m})}{v} = \frac{5.0 \times 10^4 \text{ V/m}}{v}$$

We need to obtain v before we can find B. We know that (i) a potential difference of 500 V across the plates causes the proton to deflect *vertically down* by 0.50 cm and (ii) the proton travels a horizontal distance of 5.0 cm in the same time (t) as it travels vertically down by 0.50 cm. From kinematics,

$$x_f - x_i = v_{xi}(t_f - t_i) + \tfrac{1}{2}a_x(t_f - t_i)^2 = v(t - 0 \text{ s}) \quad \Rightarrow \quad v = \frac{x_f - x_i}{t} = \frac{5.0 \text{ cm}}{t}$$

The time t can be found from the vertical motion as follows:

$$y_f - y_i = v_{yi}(t_f - t_i) + \tfrac{1}{2}a_y(t_f - t_i)^2 \quad \Rightarrow \quad 0.50 \text{ cm} = \tfrac{1}{2}a_y t^2$$

The force on the electron is

$$F_E = eE = \frac{e\Delta V}{d} = ma_y \quad \Rightarrow \quad a_y = \frac{e\Delta V}{md}$$

$$0.50 \times 10^{-2} \text{ m} = \frac{1}{2}\frac{e\Delta V}{md}t^2 = \frac{1}{2}\frac{(1.6 \times 10^{-19} \text{ C})(500 \text{ V})t^2}{(1.67 \times 10^{-27} \text{ kg})(1.0 \times 10^{-2} \text{ m})} \quad \Rightarrow \quad t = 4.569 \times 10^{-8} \text{ s}$$

The speed of the electron is

$$v = \frac{5.0 \text{ cm}}{t} = \frac{5.0 \times 10^{-2} \text{ m}}{4.569 \times 10^{-8} \text{ s}} = 1.094 \times 10^{6} \text{ m/s} \quad \Rightarrow \quad B = \frac{5.0 \times 10^{4} \text{ V/m}}{1.094 \times 10^{6} \text{ m/s}} = 46 \text{ mT}$$

Using the right hand rule we determine that this magnetic field must be directed *into the page* in order for the magnetic force to point opposite of the electric force. Thus $\vec{B} = (46 \text{ mT, into page})$.

37.33. Model: Assume that the electrons transfer all their energy to the foil.
Solve: From Chapter 17, the rise in thermal energy of the foil is

$$\Delta E_{th} = mc\Delta T = (10 \times 10^{-6} \text{ kg})(385 \text{ J/kg K})(6.0°C) = 0.0231 \text{ J}$$

This energy is provided by the electron impacts. Each electron that strikes the foil has been accelerated through a potential difference of 2000 V. Its kinetic energy when it strikes the foil is $K = e\Delta V = 2000 \text{ eV} = 3.2 \times 10^{-16}$ J. The number N of electrons that strike the foil in 10 seconds is

$$N = \frac{\Delta E_{th}}{K} = \frac{0.0231 \text{ J}}{3.2 \times 10^{-16} \text{ J}} = 7.2 \times 10^{13}$$

The current of the electron beam is

$$\frac{Nq}{\Delta t} = \frac{(7.2 \times 10^{13})(1.6 \times 10^{-19} \text{ C})}{10 \text{ s}} = 1.2 \text{ } \mu A$$

37.37. Model: (a) $V_{atom} = \frac{4\pi}{3}r_{atom}^3 = 9.05 \times 10^{-31}$ m³. Since the atomic mass of aluminum is 27 u, the mass per atom is

$$m = (27)(1.661 \times 10^{-27} \text{ kg}) = 4.48 \times 10^{-26} \text{ kg}$$

Therefore, the average density of an aluminum atom is

$$\rho_{atom} = \frac{4.48 \times 10^{-26} \text{ kg}}{9.05 \times 10^{-31} \text{ m}^3} = 4.96 \times 10^{4} \text{ kg/m}^3 \approx 5.0 \times 10^{4} \text{ kg/m}^3$$

(b) The average density of an aluminum atom is found in part (a) to be larger than the density of solid aluminum ($\rho_{Al} = 2700$ kg/m³). The volume per atom in the solid is

$$\frac{4.48 \times 10^{-26} \text{ kg}}{2700 \text{ kg/m}^3} = 1.66 \times 10^{-29} \text{ m}^3 \approx 1.7 \times 10^{-29} \text{ m}^3$$

So, using $\frac{4\pi}{3}r_{sphere}^3 = 1.66 \times 10^{-29}$ m³, we get $r_{sphere} = 1.6 \times 10^{-10}$ m.

(c) The atomic mass is almost entirely in the nucleus, so the density of the nucleus is

$$\rho_{nucleus} = \frac{4.48\times10^{-26} \text{ kg}}{\frac{4\pi}{3}(4\times10^{-15} \text{ m})^3} = 1.7\times10^{17} \text{ kg/m}^3$$

Compared to the density of solid aluminum, which is 2700 kg/m^3, the nuclear density is approximately $6.2\times10^{13} \approx 10^{14}$ times larger.

37.39. Model: The nucleus of an atom is very small and it contains protons and neutrons.
Solve: (a) The repulsive electric force between two protons in the nucleus is

$$F_E = \frac{1}{4\pi\epsilon_0} \frac{e^2}{(2.0 \text{ fm})^2} = \frac{(8.99\times10^9 \text{ N m}^2/\text{C}^2)(1.60\times10^{-19} \text{ C})^2}{(2.0\times10^{-15} \text{ m})^2} = 58 \text{ N}$$

(b) The attractive gravitational force between two protons in the nucleus is

$$F_G = \frac{Gm^2}{(2.0 \text{ fm})^2} = \frac{(6.67\times10^{-11} \text{ N m}^2/\text{kg}^2)(1.67\times10^{-27} \text{ kg})^2}{(2.0\times10^{-15} \text{ m})^2} = 4.7\times10^{-35} \text{ N}$$

Because $F_G \ll F_E$, gravitational force could not be the force to hold two protons together.

(c) The nuclear force must be very strong to overcome F_E and it must be independent of charge because both protons and neutrons are held in the nucleus very tightly. Furthermore, the nuclear force must be a very short range force since it is not felt outside the nucleus.

37.43. Model: Assume the ^{16}O nucleus is at rest. Energy is conserved.
Visualize:

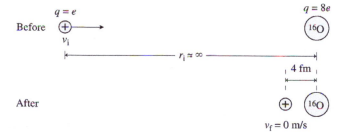

A proton with initial velocity v_i and zero electric potential energy (due to its infinite separation from an ^{16}O nucleus) moves toward ^{16}O and reverses direction when it is 1.0 fm from the surface. Since r is measured from the center of the nucleus, $r_f = 4.0$ fm.
Solve: (a) The energy conservation equation $K_f + U_f = K_i + U_i$ gives

$$0 \text{ J} + \frac{1}{4\pi\epsilon_0} \frac{e(8e)}{(3.0 \text{ fm} + 1.0 \text{ fm})} = \tfrac{1}{2}mv_i^2 + 0 \text{ J}$$

$$v_i^2 = \frac{(2)(9.0\times10^9 \text{ N m}^2/\text{C}^2)(8)(1.6\times10^{-19} \text{ C})^2}{(4.0\times10^{-15} \text{ m})(1.67\times10^{-27} \text{ kg})} = 5.52\times10^{14} \text{ m}^2/\text{s}^2 \implies v_i = 2.3\times10^7 \text{ m/s}$$

(b) $K = \tfrac{1}{2}mv_i^2 = \tfrac{1}{2}(1.67\times10^{-27} \text{ kg})(5.52\times10^{14} \text{ m}^2/\text{s}^2) = 4.60\times10^{-13} \text{ J} = 2.9 \text{ MeV}$

QUANTIZATION

Exercises and Problems

Section 38.1 The Photoelectric Effect

Section 38.2 Einstein's Explanation

38.1. Solve: A steady photoelectric current of $10\,\mu A$ is indicated in the graph. The number of electrons per second is

$$10\,\mu A = 10\,\frac{\mu C}{s} = 1.0 \times 10^{-5}\,\frac{C}{s} \times \frac{1\ \text{electron}}{1.6 \times 10^{-19}\ C} = 6.25 \times 10^{13}\ \text{electrons/s}$$

38.3. Model: Light of frequency f consists of discrete quanta, each of energy $E = hf$.
Solve: The lowest photon energy that creates photoelectrons from the metal is

$$E = \frac{hc}{\lambda} = \frac{(6.63 \times 10^{-34}\ J\ s)(3.0 \times 10^{8}\ m/s)}{388 \times 10^{-9}\ m} \times \frac{1\ eV}{1.6 \times 10^{-19}\ J} = 3.20\ eV$$

The work function of the metal is $E_0 = 3.20\ eV$.

Section 38.3 Photons

38.7. Solve: (a) The frequency of the photon is

$$f = \frac{c}{\lambda} = \frac{3.00 \times 10^{8}\ m/s}{550 \times 10^{-9}\ m} = 5.45 \times 10^{14}\ Hz$$

From Equation 38.4, the energy is

$$E = hf = (4.14 \times 10^{-15}\ eV\ s)(4.29 \times 10^{14}\ Hz) = 2.26\ eV$$

(b) The frequency of the photon is

$$f = \frac{E}{h} = \frac{7500\ eV}{4.14 \times 10^{-15}\ eV\ s} = 1.812 \times 10^{18}\ Hz$$

Thus, the wavelength is

$$\lambda = \frac{c}{f} = \frac{3.00 \times 10^{8}\ m/s}{1.208 \times 10^{18}\ Hz} = 1.656 \times 10^{-10}\ m = 0.166\ nm$$

Assess: Because x-ray photons are very energetic, their wavelength is small.

38.11. Solve: The laser light delivers 2.50×10^{17} photons per second and 100×10^{-3} J of energy per second. Thus, the energy of each photon is

$$\frac{100 \times 10^{-3} \text{ J/s}}{2.50 \times 10^{17} \text{ s}^{-1}} = 4.00 \times 10^{-19} \text{ J}$$

From Equation 38.4, the wavelength of the photons is

$$\lambda = \frac{c}{f} = \frac{hc}{E} = \frac{(6.63 \times 10^{-34} \text{ J s})(3.00 \times 10^{8} \text{ m/s})}{4.00 \times 10^{-19} \text{ J}} = 4.97 \times 10^{-7} \text{ m} = 497 \text{ nm}$$

Assess: The wavelength is in the visible region.

38.13. Model: All the light emitted by the light bulb is assumed to have a wavelength of 600 nm.
Solve: The rate of photon emission is

$$R_{\text{photon}} = \frac{P}{hf} = \frac{5 \text{ W}}{hf} = \frac{5 \text{ W}}{(6.63 \times 10^{-34} \text{ J s})(5.0 \times 10^{14} \text{ Hz})} = 1.5 \times 10^{19} \text{ photons/s} \approx 1 \times 10^{19} \text{ photons/s}$$

Section 38.4 Matter Waves and Energy Quantization

38.15. Solve: The de Broglie wavelength is $\lambda = h/mv$. Thus,

$$v = \frac{h}{m\lambda} = \frac{6.63 \times 10^{-34} \text{ J s}}{(9.11 \times 10^{-31} \text{ kg})(500 \times 10^{-9} \text{ m})} = 1456 \text{ m/s}$$

A potential difference of ΔV will raise the kinetic energy of a rest electron by $\frac{1}{2}mv^2$. Thus,

$$e\Delta V = \frac{1}{2}mv^2 \Rightarrow \Delta V = \frac{mv^2}{2e} = \frac{(9.11 \times 10^{-31} \text{ kg})(1456 \text{ m/s})^2}{2(1.6 \times 10^{-19} \text{ C})} = 6.0 \times 10^{-6} \text{ V}$$

Assess: A mere 6.0×10^{-6} V is able to increase an electron's speed to 1456 m/s.

38.21. Model: For a "particle in a box," the energy is quantized.
Solve: The energy of the $n = 1$ state is

$$E_1 = (1)^2 \frac{h^2}{8mL^2} = E_{\text{photon}} = \frac{hc}{\lambda} \Rightarrow L = \sqrt{\frac{h\lambda}{8mc}} = \sqrt{\frac{(6.63 \times 10^{-34} \text{ J s})(600 \times 10^{-9} \text{ m})}{8(9.11 \times 10^{-31} \text{ kg})(3.0 \times 10^{8} \text{ m/s})}} = 0.427 \text{ nm}$$

Section 38.5 Bohr's Model of Atomic Quantization

38.23. Model: To conserve energy, the absorption spectrum must have exactly the energy gained by the atom in the quantum jumps.
Solve: (a) An electron with a kinetic energy of 2.00 eV can collide with an atom in the $n = 1$ state and raise its energy to the $n = 2$ state. This is possible because $E_2 - E_1 = 1.50$ eV is less than 2.00 eV. On the other hand, the atom cannot be excited to the $n = 3$ state.
(b) The atom will absorb 1.50 eV of energy from the incoming electron, leaving the electron with 0.50 eV of kinetic energy.

38.25. Model: The electron must have $k \geq \Delta E_{\text{atom}}$ to cause collisional excitation. The atom is initially in the $n = 1$ ground state.

Visualize:

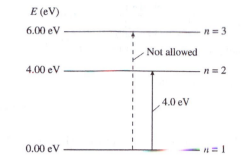

Solve: The kinetic energy of the incoming electron is

$$E = \tfrac{1}{2}mv^2 = \tfrac{1}{2}(9.11\times10^{-31}\text{ kg})(1.30\times10^6\text{ m/s})^2 = 7.698\times10^{-19}\text{ J} = 4.81\text{ eV}$$

The electron has enough energy to excite the atom to the $n=2$ stationary state $(E_2 - E_1 = 4.00\text{ eV})$. However, it does not have enough energy to excite the atom into the $n=3$ state, which requires a total energy of 6.00 eV.

Section 38.6 The Bohr Hydrogen Atom

38.29. Solve: (a) Using the data in Table 38.2, the wavelength of the electron in the $n=1$ state is

$$\lambda_1 = \frac{h}{mv} = \frac{(6.63\times10^{-34}\text{ J s})}{(9.11\times10^{-31}\text{ kg})(2.19\times10^6\text{ m/s})} = 0.332\text{ nm}$$

Likewise, $\lambda_2 = 0.665$ nm, and $\lambda_3 = 0.997$ nm.

(b) For $n=1$, the circumference of the orbit is $0.0529\text{ nm}\times2\pi = 0.332$ nm, which is exactly equal to λ_1. For $n=2$, the circumference of the orbit is

$$0.212\text{ nm}\times2\pi = 1.332\text{ nm} = 2\lambda_2 = 2(0.665\text{ nm})$$

Likewise, the data from part (a) and Table 38.2 shows $3\lambda_3 = 2\pi r_3$.

(c)

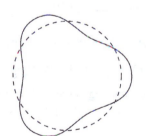

Section 38.7 The Hydrogen Spectrum

38.33. Model: Equation 38.37 predicts the absorption spectrum of hydrogen if we let $m=1$.

$$\lambda_{1\to n} = \frac{91.18\text{ nm}}{\left(\frac{1}{1^2}-\frac{1}{n^2}\right)} \qquad n = 2, 3, 4, \ldots$$

Visualize: The $1\to2$ transition will produce the longest wavelength; the $1\to3$ transition will produce the second-longest wavelength; and $1\to4$ transition will produce the third-longest wavelength.

Solve:

$$\lambda_{1 \to 4} = \frac{91.18 \text{ nm}}{\left(\frac{1}{1^2} - \frac{1}{4^2}\right)} = 97.26 \text{ nm}$$

Assess: This result is in the ultraviolet region, as expected for $m = 1$.

38.35. Solve: For hydrogen-like ions, Equation 38.38 is

$$r_n = n^2 \frac{a_B}{Z} = \frac{(r_n)_H}{Z} \qquad v_n = Z\frac{v_1}{n} = Z \cdot (v_n)_H \qquad E_n = -Z^2\left(\frac{13.60 \text{ eV}}{n^2}\right) = Z^2(E_n)_H$$

Where $(r_n)_H$, $(v_n)_H$, and $(E_n)_H$ are the values of ordinary hydrogen. He^+ has $Z = 2$. Using Table 38.2 for the values of hydrogen, we get

n	r_n(nm)	v_n(m/s)	E_n(eV)
1	0.026	4.38×10^6	-54.4
2	0.106	2.19×10^6	-13.6
3	0.238	1.46×10^6	-6.0

38.37. Visualize: Equation 38.8 gives V_{stop} in terms of f; we replace f with c/λ.

$$V_{stop} = \frac{hc/\lambda - E_0}{e}$$

The value of E_0 for aluminum is found in Table 38.1: $E_0 = 4.28 \text{ eV} = 6.848 \times 10^{-19}$ J.

Solve: For $\lambda = 250$ nm:

$$V_{stop} = \frac{hc/\lambda - E_0}{e} = \frac{(6.63 \times 10^{-34} \text{ J} \cdot \text{s})(3.00 \times 10^8 \text{ m/s})/(250 \times 10^{-9} \text{ m}) - 6.848 \times 10^{-19} \text{ J}}{1.60 \times 10^{-19} \text{ C}} = 0.693 \text{ V}$$

For $\lambda = 200$ nm:

$$V_{stop} = \frac{hc/\lambda - E_0}{e} = \frac{(6.63 \times 10^{-34} \text{ J} \cdot \text{s})(3.00 \times 10^8 \text{ m/s})/(200 \times 10^{-9} \text{ m}) - 6.848 \times 10^{-19} \text{ J}}{1.60 \times 10^{-19} \text{ C}} = 1.94 \text{ V}$$

So the change in stopping potential is $1.94 \text{ V} - 0.693 \text{ V} = 1.24 \text{ V}$.

Assess: This answer looks reasonable in light of Figure 38.10.

38.41. Solve: (a) The maximum kinetic energy of photoelectrons is $K_{max} = hf - E_0$. Substituting the given values,

$$2.8 \text{ eV} = \frac{hc}{\lambda} - E_0 \quad 1.1 \text{ eV} = \frac{hc}{1.5\lambda} - E_0$$

Multiplying the second equation by 1.5 and subtracting the second equation from the first,

$$1.15 \text{ eV} = 0.5 \, E_0 \Rightarrow E_0 = 2.3 \text{ eV}$$

(b) Substituting $E_0 = 2.3$ eV into the first equation,

$$2.8 \text{ eV} = \frac{(4.14 \times 10^{-15} \text{ eV s})(3.0 \times 10^8 \text{ m/s})}{\lambda} - 2.3 \text{ eV} \Rightarrow \lambda = 244 \text{ nm}$$

38.43. Solve: (a) The stopping potential is

$$V_{stop} = \frac{h}{e}f - \frac{h}{e}f_0$$

A graph of V_{stop} versus frequency f should be linear with x-intercept f_0 and slope h/e. Since the x-intercept is $f_0 = 1.0 \times 10^{15}$ Hz, the work function is

$$E_0 = hf_0 = (4.14 \times 10^{-15} \text{ eV s})(1.0 \times 10^{15} \text{ Hz}) = 4.14 \text{ eV}$$

(b) The slope of the graph is

$$\frac{\Delta V_{stop}}{\Delta f} = \frac{8.0 \text{ V} - 0 \text{ V}}{3.0 \times 10^{15} \text{ Hz} - 1.0 \times 10^{15} \text{ Hz}} = 4.0 \times 10^{-15} \text{ V s}$$

Because the slope of the V_{stop} versus f graph is h/e, an experimental value of Planck's constant is

$$h = e(4.0 \times 10^{-15} \text{ V s}) = (1.6 \times 10^{-19} \text{ C})(4.0 \times 10^{-15} \text{ V s}) = 6.4 \times 10^{-34} \text{ J s}$$

Assess: This value of the Planck's constant is about 3.5% lower than the accepted value.

38.47. Visualize: The mass of the cell is $m = \rho V = (1100 \text{ kg/m}^3)[\pi(3.5 \ \mu m)^2(2.0 \ \mu m)] = 8.47 \times 10^{-14} \text{ kg}$.
Solve: The de Broglie wavelength is

$$\lambda = \frac{h}{mv} = \frac{6.63 \times 10^{-34} \text{ J s}}{(8.47 \times 10^{-14} \text{ kg})(4.0 \text{ mm/s})} = 2.0 \times 10^{-18} \text{ m}$$

This wavelength is so tiny that we do not need to be concerned with the wave nature of blood cells when describing blood flow.
Assess: We seem to have gotten along just fine without considering the wave nature of blood cells before.

38.49. Model: Use the wave model for neutrons; they will interfere as they travel through the two-slit apparatus.
Visualize: From the prior chapter on the wave model of light we know that the fringe spacing in a two-slit experiment with slit spacing d and screen distance L is $\Delta y = \dfrac{\lambda L}{d}$. Using the scale on the figure, the five tallest peaks, which are $4\Delta y$, appear to span about 250 μm, so $\Delta y = 250 \ \mu m = 4 \approx 63 \ \mu m$. We are given $d = 0.10$ mm and $L = 3.5$ m.
Solve: Solve the de Broglie wavelength $\lambda = \dfrac{h}{mv}$ for v.

$$v = \frac{h}{m\lambda} = \frac{h}{m\left(\dfrac{d}{L}\Delta y\right)} = \frac{6.63 \times 10^{-34} \text{ J s}}{(1.67 \times 10^{-27} \text{ kg})\left[\dfrac{0.10 \text{ mm}}{(3.5 \text{ m})}(63 \ \mu m)\right]} = 220 \text{ m/s} \approx 200 \text{ m/s}$$

Because of the estimation in reading the fringe spacing from the figure, we round the answer to one significant figure.
Assess: This speed seems reasonable for neutrons.

38.53. Model: The energy of the emitted gamma-ray photon is exactly equal to the energy between levels 1 and 2.
Solve: From Equation 38.15, the energy levels of the proton are

$$E_n = n^2 \frac{h^2}{8 mL^2} \quad n = 1, 2, 3, \ldots$$

The energy of the emitted photon is

$$E_2 - E_1 = \frac{4h^2}{8mL^2} - \frac{h^2}{8mL^2} = \frac{3h^2}{8mL^2}$$

$$\Rightarrow L = \sqrt{\frac{3h^2}{8m(E_2 - E_1)}} = \sqrt{\frac{3(6.63 \times 10^{-34} \text{ J s})^2}{8(1.67 \times 10^{-27} \text{ kg})(2.0 \times 10^6 \text{ eV})} \times \frac{1 \text{ eV}}{1.60 \times 10^{-19} \text{ J}}} = 1.8 \times 10^{-14} \text{ m} = 18 \text{ fm}$$

Assess: This is roughly the size of a typical nucleus.

38.57. Solve: A photon with wavelength $\lambda = 1240$ nm has an energy $E_{photon} = hf = hc/\lambda = 1.0$ eV. Because E_{photon} must exactly match ΔE of the atom, a 1240 nm photon can be emitted only in a $3 \rightarrow 2$ transition. So, after the collision the atom was in the $n = 3$ state. Before the collision, the atom was in its ground state ($n = 1$). Thus, an electron with $v_i = 1.4 \times 10^6$ m/s collided with the atom in the $n = 1$ state. The atom gained 4.5 eV in the collision as it was excited from the $n = 1$ to $n = 3$, so the electron lost 4.5 eV $= 7.20 \times 10^{-19}$ J of kinetic energy. Initially, the kinetic energy of the electron was

$$K_i = \tfrac{1}{2}m_{elec}v_i^2 = \tfrac{1}{2}(9.11 \times 10^{-31} \text{ kg})(1.40 \times 10^6 \text{ m/s})^2 = 8.93 \times 10^{-19} \text{ J}$$

After losing 7.20×10^{-19} J in the collision, the kinetic energy is

$$K_f = K_i - 7.20 \times 10^{-19} \text{ J} = 1.73 \times 10^{-19} \text{ J} = \tfrac{1}{2} m_{elec} v_f^2 \Rightarrow v_f = \sqrt{\frac{2K_f}{m_{elec}}} = \sqrt{\frac{2(1.73 \times 10^{-19} \text{ J})}{9.11 \times 10^{-31} \text{ kg}}} = 6.2 \times 10^5 \text{ m/s}$$

38.59. Solve: The wavelengths in the hydrogen spectrum are given by

$$\lambda_{n \to m} = \frac{91.18 \text{ nm}}{1/m^2 - 1/n^2}$$

For each m, the wavelengths range from a maximum value when $n = m+1$ to a minimum value (the series limit) when $n \to \infty$. A few calculations reveal the following behavior:

m	λ_{max} for $n = m+1$	λ_{min} for $n = \infty$
1	122 nm	91 nm
2	656 nm	365 nm
3	1876 nm	822 nm
4	4050 nm	1459 nm

We see that the *only* visible wavelengths (400–700 nm) in the hydrogen spectrum occur for $m = 2$, starting with $n = 3$. We can calculate that $\lambda_{3 \to 2} = 656.5$ nm, $\lambda_{4 \to 2} = 486.3$ nm, $\lambda_{5 \to 2} = 434.2$ nm, and $\lambda_{6 \to 2} = 410.3$ nm. The next transition, $6 \to 2$, has $\lambda < 400$ nm. Thus these are the only four visible wavelengths in the hydrogen spectrum.

38.61. Solve: (a) From Equation 38.37, the wavelengths of the emission spectrum are

$$\lambda_{n \to m} = \frac{91.18 \text{ nm}}{m^{-2} - n^{-2}} \quad m = 1, 2, 3, \dots \quad n = m+1, m+2, \dots$$

For the $200 \to 199$ transition,

$$\lambda_{200 \to 199} = \frac{91.18 \text{ nm}}{(199)^{-2} - (200)^{-2}} = 0.362 \text{ m}$$

(b) For the $2 \to 199$ transition,

$$\lambda_{2 \to 199} = \frac{91.18 \text{ nm}}{(2)^{-2} - (199)^{-2}} = 4.000404071 \ (91.18 \text{ nm})$$

Likewise, $\lambda_{2 \to 200} = 4.00040004(91.18 \text{ nm})$. The difference in the wavelengths of these two transitions is $0.0000041 \times$ $(91.18 \text{ nm}) = 0.000368$ nm.

38.63. Visualize: A neutral oxygen atom that has lost 7 electrons is a hydrogen-like atom because it has one electron going around a nucleus with $Z = 8$.

Solve: The energy levels for the O^{+7} ion are

$$E_n = -\frac{13.60 \, Z^2 \text{ eV}}{n^2} = -\frac{13.60(8)^2 \text{ eV}}{n^2} = -\frac{870.4 \text{ eV}}{n^2}$$

The energy of the $3 \to 2$ transition is

$$E_3 - E_2 = -870.4 \text{ eV}\left(\frac{1}{3^2} - \frac{1}{2^2}\right) = 120.89 \text{ eV} = \frac{hc}{\lambda}$$

$$\Rightarrow \lambda = \frac{(4.14 \times 10^{-15} \text{ eV})(3.0 \times 10^8 \text{ m/s})}{120.89 \text{ eV}} = 10.28 \text{ nm}$$

Likewise for the $4 \to 2$ transition,

$$E_4 - E_2 = -870.4 \text{ eV}\left(\frac{1}{4^2} - \frac{1}{2^2}\right) = 163.2 \text{ eV}$$

The wavelength for this transition $\lambda = 7.62$ nm. For the $5 \to 2$ transition, $E_5 - E_2 = 182.78$ eV and $\lambda = 6.80$ nm. All the wavelengths are in the ultraviolet range.

WAVE FUNCTIONS AND UNCERTAINTY

Exercises and Problems

Section 39.1 Waves, Particles, and the Double-Slit Experiment

39.1. Model: The sum of the probabilities of all possible outcomes must equal 1 (100%).
Solve: The sum of the probabilities is $P_A + P_B + P_C + P_D = 1$. Hence,

$$0.40 + 0.30 + P_C + P_D = 1 \implies P_C + P_D = 0.30$$

Because $P_C = 2P_D$, $2P_D + P_D = 0.30$. This means $P_D = 0.10$ and $P_C = 0.20$. Thus, the probabilities of outcomes C and D are 20% and 10%, respectively.

39.3. Model: The probability that the outcome will be A or B is the sum of P_A and P_B.
Solve: (a) A regular deck of cards has 52 cards. Drawing a given card (say, queen of hearts) from this deck has a probability of 1/52. Because there are 4 aces in the deck, the probability of drawing an ace is 1/52 + 1/52 + 1/52 + 1/52 = 4(1/42) = 4/52 = 0.077 = 7.7%.
(b) Because there are 13 spades, the probability of drawing a spade is 13(1/52) = 13/52 = 0.25 = 25%.

39.2 Connecting the Wave and Photon Views

39.9. Model: See Example 39.1.
Solve: We are given $N_{tot} = 5.0 \times 10^{12}$, $N(\text{in } \delta x \text{ at } x) = 2.0 \times 10^9$, and $\delta x = 0.10$ mm. The probability that a photon lands on the strip is $\text{Prob(in } \delta x \text{ at } x) = 2.0 \times 10^9 / 5.0 \times 10^{12} = 0.00040$. Solve for the probability density $P(x)$ in Equation 39.12.

$$P(x) = \frac{\text{Prob(in } \delta x \text{ at } x)}{\delta x} = \frac{0.00040}{1.0 \times 10^{-4} \text{ m}} = 4.0 \text{ m}^{-1}$$

Assess: This result is similar to the result of Example 39.1.

Section 39.3 The Wave Function

39.11. Model: The probability of finding a particle at position x is determined by $|\psi(x)|^2$.

Solve: (a) The probability of detecting an electron is $\text{Prob(in } \delta x \text{ at } x) = |\psi(x)|^2 \delta x$. At $x = 0.000$ mm, the number of electrons landing in a 0.010-mm-wide strip is

$$\frac{N}{N_{\text{total}}} = |\psi(0 \text{ mm})|^2 \delta x \implies N = |\psi(0 \text{ mm})|^2 \delta x N_{\text{total}} = \left(\tfrac{1}{3} \text{ mm}^{-1}\right)(0.010 \text{ mm})(1.0 \times 10^6) = 3333 \approx 3.3 \times 10^3$$

(b) Likewise, the number of electrons landing in a 0.010-mm-wide strip at $x = 2.000$ mm is

$$N = |\psi(2.0 \text{ mm})|^2 \delta x N_{\text{total}} = (0.111 \text{ mm}^{-1})(0.010 \text{ mm})(1.0 \times 10^6) = 1111 \approx 1.1 \times 10^3$$

39.13. Model: The probability of finding a particle at position x is determined by $|\psi(x)|^2$.

Solve: (a) The probability of detecting an electron is $\text{Prob(in } \delta x \text{ at } x) = |\psi(x)|^2 \, \delta x$. Hence the probability the electron will land in a $\delta x = 0.010$-mm-wide strip at $x = 0.000$ mm is

$$\text{Prob(in } 0.010 \text{ mm at } x = 0.000 \text{ mm)} = (0.50 \text{ mm}^{-1})(0.010 \text{ mm}) = 5.0 \times 10^{-3}$$

(b) Since $|\psi(0.500 \text{ mm})|^2 = 0.25 \text{ mm}^{-1}$, the probability is 2.5×10^{-3}.

(c) Since $|\psi(1.000 \text{ mm})|^2 = 0.0 \text{ mm}^{-1}$, the probability is 0.0.

(d) Since $|\psi(2.000 \text{ mm})|^2 = 0.25 \text{ mm}^{-1}$, the probability is 2.5×10^{-3}.

Section 39.4 Normalization

39.17. Model: The probability of finding the particle is determined by the probability density $P(x) = |\psi(x)|^2$.

Solve: (a) According to the normalization condition, $\int_{-\infty}^{\infty} |\psi(x)|^2 \, dx = 1$. From the given $\psi(x)$-versus-x graph, we first generate a $|\psi(x)|^2$-versus-x graph and then find the area under the curve.

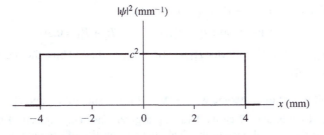

The area under the $|\psi(x)|^2$-versus-x graph is

$$\int_{-4.0 \text{ mm}}^{4.0 \text{ mm}} c^2 dx = (8.0 \text{ mm})c^2 = 1.0 \quad \Rightarrow \quad c = \sqrt{\frac{1.0}{8.0 \text{ mm}}} = 0.35 \text{ mm}^{-1/2}$$

(b) The graph is shown above.
(c) The probability is

$$\text{Prob(1.0 mm } \leq x \leq 1.0 \text{ mm)} = \int_{-1.0 \text{ mm}}^{1.0 \text{ mm}} c^2 dx = c^2 (2.0 \text{ mm}) = \left(\frac{1.0}{8.0}\text{mm}^{-1}\right)(2.0 \text{ mm}) = 0.25$$

Section 39.5 Wave Packets

39.21. Model: A laser pulse is an electromagnetic wave packet, hence it must satisfy the relationship $\Delta f \, \Delta t \approx 1$.
Solve: Because $c = \lambda f$, the frequency and period are

$$f = \frac{3.0 \times 10^8 \text{ m/s}}{1.5 \times 10^{-6} \text{ m}} = 2.0 \times 10^{14} \text{ Hz} \quad \Rightarrow \quad T = \frac{1}{f} = \frac{1}{2.0 \times 10^{14} \text{ Hz}} = 5.0 \times 10^{-15} \text{ s}$$

Since $\Delta f = 2.0$ GHz, the minimum pulse duration is

$$\Delta t \approx \frac{1}{\Delta f} = \frac{1}{2.0 \times 10^9 \text{ Hz}} = 5.0 \times 10^{-10} \text{ s}$$

The number n of oscillations in this laser pulse is

$$n = \frac{\Delta t}{T} = \frac{5.0 \times 10^{-10} \text{ s}}{5.0 \times 10^{-15} \text{ s}} = 1.0 \times 10^5 \text{ oscillations}$$

Section 39.6 The Heisenberg Uncertainty Principle

39.23. Model: Andrea is subject to the Heisenberg uncertainty principle. She cannot be absolutely at rest ($\Delta v_x = 0$) without violating the uncertainty principle. Andrea's mass is 50 kg.
Solve: Because Andrea is inside her room, her position uncertainty could be as large as $\Delta x = L = 5.0$ m. According to the uncertainty principle, her velocity uncertainty is thus

$$\Delta v_x = \frac{h}{2m\,\Delta x} = \frac{6.63\times10^{-34}\ \text{J s}}{2(50\ \text{kg})(5.0\ \text{m})} = 1.3\times10^{-36}\ \text{m}$$

Because her *average* velocity is zero, her velocity is likely to be in the range -0.65×10^{-36} m/s to $+0.65\times10^{-36}$ m/s.

Assess: At a speed of 1×10^{-36} m/s, Andrea would have moved less than 1% the diameter of the nucleus of an atom in the entire age of the universe. It makes perfect sense to think that macroscopic objects can be at rest.

39.25. Model: Protons are subject to the Heisenberg uncertainty principle.
Solve: We know the proton is somewhere within the nucleus, so the uncertainty in our knowledge of its position is at most $\Delta x = L = 4.0$ fm. With a finite Δx, the uncertainty Δp_x is given by the uncertainty principle:

$$\Delta p_x = m\Delta v_x = \frac{h/2}{\Delta x} \quad\Rightarrow\quad \Delta v_x = \frac{h}{2mL} = \frac{6.63\times10^{-34}\ \text{J s}}{2(1.67\times10^{-27}\ \text{kg})(4.0\times10^{-15}\ \text{m})} = 5.0\times10^{7}\ \text{m/s}$$

Because the average velocity is zero, the best we can say is that the proton's velocity is somewhere in the range -2.5×10^{7} m/s to 2.5×10^{7} m/s. Thus, the smallest range of speeds is 0 to 2.5×10^{7} m/s.

39.27. Model: A radar pulse comprised of electromagnetic waves is a wave packet, so it must satisfy the relationship $\Delta f\,\Delta t \approx 1$.
Solve: The period 0.100 ns for a wave corresponds to a frequency of

$$f = \frac{1}{T} = \frac{1}{0.100\ \text{ns}} = 10.0\times10^{9}\ \text{Hz}$$

A pulse of duration $\Delta t = 1.0$ ns is 10 oscillations of the wave. Although the station broadcasts at a center frequency of 1.0×10^{10} Hz, this pulse is not a pure 1.0×10^{10} Hz oscillation. Instead, this pulse has been created by the superposition of many waves whose frequencies span

$$\Delta f \approx \frac{1}{\Delta t} = \frac{1}{1.0\times10^{-9}\ \text{s}} = 1.0\times10^{9}\ \text{Hz}$$

This range of frequencies is centered at the 1.0×10^{10} Hz broadcast frequency, so the waves that must be superimposed to create this pulse span the frequency range:

$$f - \Delta f/2 < f < f + \Delta f/2 \quad\Rightarrow\quad 9.5\ \text{GHz} < f < 10.5\ \text{GHz}$$

39.31. Model: The probability of finding a particle at position x is determined by $P(x) = |\psi(x)|^2$.
Visualize:

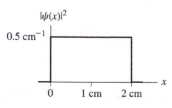

Solve: (a) Since the electrons are uniformly distributed over the interval $0 \le x \le 2$ cm, the probability density $P(x) = |\psi(x)|^2$ is constant over this interval. $P(x) = 0$ outside this interval because no electrons are detected. Thus $|\psi(x)|^2$ is a square function, as shown in the figure. To be normalized, the area under the probability curves must be 1. Hence, the peak value of $|\psi(x)|^2$ must be $0.5\ \text{cm}^{-1}$.

(b) The interval is $\delta x = 0.02$ cm. The probability is

$$\text{Prob(in } \delta x \text{ at } x = 0.80 \text{ cm)} = |\psi(x = 0.80 \text{ cm})|^2 \, \delta x = (0.5 \text{ cm}^{-1})(0.02 \text{ cm}) = 0.01 = 1\%$$

(c) From Equation 39.7, the number N of electrons that will fall within the given interval is

$$N(\text{in } \delta x \text{ at } x = 0.80 \text{ cm}) = N_{\text{total}}\text{Prob(in } dx \text{ at } x = 0.80 \text{ cm)} = 10^6 \times (0.01) = 1 \times 10^4$$

(d) The probability density is $P(x = 0.80 \text{ cm}) = |\psi(x = 0.80 \text{ cm})|^2 = 0.5 \text{ cm}^{-1}$.

39.33. Model: The probability of finding a particle at position x is determined by $P(x) = |\psi(x)|^2$.

Visualize:

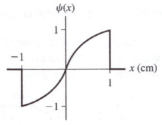

Solve: (a) Yes, because the area under the $|\psi(x)|^2$ curve is equal to 1.

(b) There are two things to consider when drawing $\psi(x)$. First $\psi(x)$ is an *oscillatory* function that changes sign every time it reaches zero. Second, $\psi(x)$ must have the right shape. Each point on the $\psi(x)$ curve is the square root of the corresponding point on the $|\psi(x)|^2$ curve. The values $|\psi(x)|^2 = 1 \text{ cm}^{-1}$ and $|\psi(x)|^2 = 0 \text{ cm}^{-1}$ clearly give $\psi(x) = \pm 1 \text{ cm}^{-1/2}$ and $\psi(x) = 0 \text{ cm}^{-1/2}$, respectively. But consider $x = 0.5$ cm, where $|\psi(x)|^2 = 0.5 \text{ cm}^{-1}$. Because $\sqrt{0.5} = \pm 0.707$, $\psi(x = 0.5 \text{ cm}) = \pm 0.707 \text{ cm}^{-1/2}$. This tells us that the $\psi(x)$ curve is not linear but *bows upward* (or downward if we take the negative square root) over the interval $0 \leq x \leq 1$ cm. Thus, $\psi(x)$ has the shape shown in the above figure.

(c) $\delta x = 0.0010$ cm is a very small interval, so we can use Prob(in δx at x) $= |\psi(x)|^2 \, \delta x$. The values of $|\psi(x)|^2$ can be read from Figure P39.33. Thus,

$$\text{Prob(in } \delta x \text{ at } x = 0.00 \text{ cm)} = |\psi(x = 0.00 \text{ cm})|^2 \, \delta x = (0.00 \text{ cm}^{-1})(0.0010 \text{ cm}) = 0.00$$

$$\text{Prob(in } \delta x \text{ at } x = 0.50 \text{ cm)} = |\psi(x = 0.50 \text{ cm})|^2 \, \delta x = (0.50 \text{ cm}^{-1})(0.0010 \text{ cm}) = 0.00050$$

$$\text{Prob(in } \delta x \text{ at } x = 0.999 \text{ cm)} = |\psi(x = 0.999 \text{ cm})|^2 \, \delta x = (0.999 \text{ cm}^{-1})(0.0010 \text{ cm}) = 0.00999 \approx 0.0010$$

(d) The number N of electrons expected to land in the interval $-0.3 \text{ cm} \leq x \leq 0.3$ cm is

$$N(\text{in} - 0.30 \text{ cm} \leq x \leq 0.30 \text{ cm}) = N_{\text{total}}\text{Prob(in} - 0.30 \text{ cm} \leq x \leq 0.30 \text{ cm})$$

$$= (1 \times 10^4) \int_{-0.30 \text{ cm}}^{0.30 \text{ cm}} |\psi(x)|^2 \, dx = (1 \times 10^4)\left[2 \times \left(\tfrac{1}{2} \times 0.30 \text{ cm} \times 0.30 \text{ cm}^{-1}\right)\right]$$

$$= 900$$

39.39. Model: The probability of finding a particle at position x is determined by the probability density $P(x) = |\psi(x)|^2$.

Solve: (a) Normalization of the wave function requires that $\int_{-\infty}^{\infty} |\psi(x)|^2 \, dx = 1$. Therefore,

$$1 = c^2 \int_{-1}^{1}(1 - x^2) \, dx = c^2\left[x - \frac{x^3}{3}\right]_{-1}^{1} = \frac{4}{3}c^2 \quad \Rightarrow \quad c = \sqrt{\frac{3}{4}} = 0.87 \text{ cm}^{-1/2}$$

(b) The value of $\psi(x)$ decreases from 0.87 at $x = 0$ cm to 0.75 at $x = 0.5$ cm and then to 0 cm at $x = 1$ cm. Thus, the graph is bowed *upward* over the interval $0 \leq x \leq 1$ cm. $\psi(x) = 0$ for $x > 1$ cm. The graph is also symmetrical about $x = 0$ cm.

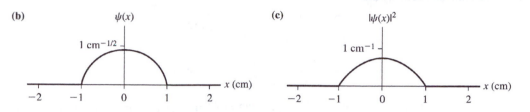

(c) The probability density is $|\psi(x)|^2 = c^2(1 - x^2) = \frac{3}{4}(1 - x^2)$. The value of $|\psi(x)|^2$ decreases from 0.75 at $x = 0$ cm to 0.56 at $x = 0.5$ cm and then to 0 at $x = 1$ cm. This graph is also bowed upward, although not as sharply as $\psi(x)$. The graph of $|\psi(x)|^2$ is shown in the above figure.

(d) The number of electrons is

$$N(\text{in } 0.00 \text{ cm } \leq x \leq 0.50 \text{ cm}) = N_{\text{total}} \text{Prob(in } 0.00 \text{ cm } \leq x \leq 0.50 \text{ cm})$$

$$\text{Prob(in } 0.00 \text{ cm } \leq x \leq 0.50 \text{ cm}) = \int_{0.00 \text{ cm}}^{0.50 \text{ cm}} |\psi(x)|^2 \, dx = \frac{3}{4} \int_{0.00}^{0.50} (1 - x^2 dx) = \frac{3}{4}\left[x - \frac{x^3}{3} \right]_{0.00}^{0.50} = 0.344$$

Thus, the number of electrons detected in the interval $0 \text{ cm} \leq x \leq 0.5 \text{ cm}$ is $10{,}000 \times 0.344 = 3440 \approx 3.4 \times 10^3$.

39.41. Model: The probability of finding a particle at position x is determined by the probability density $P(x) = |\psi(x)|^2$.

Solve: (a) The given probability density means that $\psi_1(x) = \sqrt{a/(1 - x)}$ in the range $-1 \text{ mm} \leq x \leq 0$ mm and $\psi_2(x) = \sqrt{b(1 - x)}$ in the range $0 \text{ mm} \leq x \leq 1$ mm. Because $\psi_1(x = 0 \text{ mm}) = \psi_2(x = 0 \text{ mm})$, $\sqrt{a} = \sqrt{b}$ and thus $a = b$.

(b) At $x = -1$ mm, $P(x) = \frac{1}{2}a$; at $x = 0$ mm, $P(x) = a$; and at $x = 1$ mm, $P(x) = 0$. Furthermore, $P(x)$ is a linear function of x for $0 \leq x \leq 1$ mm.

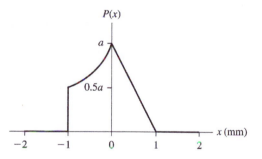

(c) Normalization of the wave function requires integrating over the entire range. This gives

$$\int_{-\infty}^{\infty} P(x)dx = \int_{-1}^{0}\left(\frac{a}{1 - x} \right)dx + \int_{0}^{1} b(1 - x)dx = -a\left[\ln(1 + x) \right]_{-1}^{0} + b\left[x - \frac{x^2}{2} \right]_{0}^{1} = a\ln 2 + \frac{b}{2} = a\ln 2 + \frac{a}{2} = a\left(\ln 2 + \frac{1}{2} \right)$$

where we have used $a = b$ from part (a). Because $\int P(x)dx = 1$, we get $a = b = 1/\left(\ln 2 + \frac{1}{2} \right) = 1/1.193 = 0.84$.

(d) The probability that the particle will be found to the left of the origin is

$$\int_{-1}^{0} P(x)dx = \int_{-1}^{0} \frac{a}{1 - x}dx = -a\left[\ln(1 - x) \right]_{-1}^{0} = a\ln 2 = \frac{1}{1.193}(0.693) = 0.581 \approx 58\%$$

39.45. Model: The electron is subject to the Heisenberg uncertainty principle. Model the nucleus as a one-dimensional box.
Solve: (a) The uncertainty in our knowledge of the position of the electron is $\Delta x = 10$ fm. The uncertainty in the electron's momentum and velocity are

$$\Delta p_x = \frac{h/2}{\Delta x} = \frac{6.63 \times 10^{-34} \text{ J s}}{2(10 \times 10^{-15} \text{ m})} = 3.32 \times 10^{-20} \text{ kg m/s}$$

$$\Delta v_x = \frac{\Delta p_x}{m} = \frac{3.32 \times 10^{-20} \text{ kg m/s}}{9.11 \times 10^{-31} \text{ kg}} = 3.64 \times 10^{10} \text{ m/s}$$

The range of possible velocities is -1.82×10^{10} m/s to $+1.82 \times 10^{10}$ m/s, so the range of speeds is from 0 m/s to 1.82×10^{10} m/s.

(b) The minimum range of speeds for an electron confined to a nucleus exceeds the speed of light, so it is not possible.

ONE-DIMENSIONAL QUANTUM MECHANICS

Exercises and Problems

Sections 40.3–40.4 A Particle in a Rigid Box

40.3. Model: Model the electron as a particle in a rigid one-dimensional box of length L.
Solve: The energy levels for a particle in a rigid box are

$$E_n = n^2 \frac{\pi^2 \hbar^2}{2L^2}$$

The wave function shown in Figure EX40.3 corresponds to $n = 4$. Thus,

$$L = \frac{4\pi\hbar}{\sqrt{2mE_4}} = \frac{4h}{2\sqrt{2mE_4}} = \frac{4(6.63\times10^{-34}\text{ J s})}{2\sqrt{2(9.11\times10^{-31}\text{ kg})(6.0\text{ eV}\times1.6\times10^{-19}\text{ J/eV})}} = 1.0\text{ nm}$$

Section 40.6 Finite Potential Wells

40.5. Solve: From Equation 40.41, the units of the penetration distance are

$$\eta = \frac{\hbar}{\sqrt{2m(U_0 - E)}} \Rightarrow \frac{\text{J}\times\text{s}}{\sqrt{\text{kg}\times\text{J}}} = \frac{(\text{kg}\times\text{m}^2/\text{s}^2)\times\text{s}}{\sqrt{\text{kg}\times\text{kg m}^2/\text{s}^2}} = \frac{\text{kg}\times\text{m}^2/\text{s}}{\sqrt{\text{kg}^2\times\text{m}^2/\text{s}^2}} = \frac{\text{kg}\times\text{m}^2/\text{s}}{\text{kg}\times\text{m/s}} = \text{m}$$

40.7. Solve: (a) According to Equation 40.41, the penetration distance is

$$\eta = \frac{\hbar}{\sqrt{2m(U_0 - E)}} = \frac{1.05\times10^{-34}\text{ J s}}{\sqrt{2(9.11\times10^{-31}\text{ kg})(2.0\text{ eV} - 0.5\text{ eV})\times1.60\times10^{-19}\text{ J/eV}}} = 0.159\text{ nm}$$

(b) Likewise for $E = 1.00$ eV, $\eta = 0.195$ nm.
(c) For $E = 1.50$ eV, $\eta = 0.275$ nm.
Assess: These values are of the correct order of magnitude as you can see by referring to Figure 40.14a.

40.9. Model: The electron is in a finite potential well whose energies and wave functions were shown in Figure 40.14a.
Visualize: We are given $U_0 = 2.00$ eV and $E = 1.50$ eV. We seek $x - L$.
Solve: First use the equation for the penetration distance.

$$\eta = \frac{\hbar}{\sqrt{2m(U_0 - E)}} = \frac{1.05\times10^{-34}\text{ J s}}{\sqrt{2(9.11\times10^{-31}\text{ kg})(0.50\text{ eV})\left(\frac{1.60\times10^{-19}\text{ J}}{1\text{ eV}}\right)}} = 0.275\text{ nm}$$

Now use this value of η in $\psi(x) = \psi_{edge}e^{-(x-L)/\eta}$, where $\psi(x)/\psi_{edge} = 0.25$ and solve for $x - L$.

$$0.25 = e^{-(x-L)/\eta} \Rightarrow \ln 0.25 = -(x-L)/\eta \Rightarrow$$
$$x - L = -\eta(\ln 0.25) = x - L = -(0.275 \text{ nm})(\ln 0.25) = 0.38 \text{ nm}$$

Assess: The penetration distance is the distance into the classically forbidden region where $\psi(x)/\psi_{edge} = 0.37$ so a little further into the forbidden region is where $\psi(x)/\psi_{edge} = 0.25$.

Section 40.7 Wave-Function Shapes

40.13. Visualize:

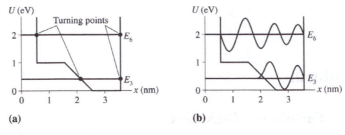

(a)   (b)

Solve: **(a)** The energy diagram is shown above.
(b) There are three factors to consider. First, the de Broglie wavelength increases as the particle's speed and kinetic energy decreases. Thus, the spacing between the nodes of $\psi(x)$ increases in regions where U is larger. Second, a particle is more likely to be found where it is moving the slowest. Thus, the amplitude of $\psi(x)$ increases in regions where U is larger. Third, $n = 3$ has 3 antinodes and $n = 6$ has six antinodes.

Section 40.8 The Quantum Harmonic Oscillator

40.17. Model: The electron is a quantum harmonic oscillator.
Solve: The energy levels of the quantum harmonic oscillator are $E_n = (n - \frac{1}{2})\hbar\omega$. The longest wavelength of light that can be absorbed is in a transition from level n to the next level $n+1$. Thus

$$E_{photon} = hf = \frac{hc}{\lambda} = \Delta E_{electron} = E_{n+1} - E_n = \hbar\omega = \frac{h}{2\pi}\sqrt{\frac{k}{m}}$$

$$\Rightarrow \lambda = 2\pi c\sqrt{\frac{m}{k}} = 2\pi(3.0\times10^8 \text{ m/s})\sqrt{\frac{9.11\times10^{-31} \text{ kg}}{12.0 \text{ N/m}}} = 5.19\times10^{-7} \text{ m} = 519 \text{ nm}$$

40.19. Model: The electron is a quantum harmonic oscillator. The given levels are adjacent levels.
Solve: Let the two adjacent levels be n and $n+1$. From Equation 40.48,

$$E_n = \left(n - \frac{1}{2}\right)\hbar\omega = 2.0 \text{ eV} \qquad E_{n+1} = \left(n+1-\frac{1}{2}\right)\hbar\omega = 2.8 \text{ eV}$$

$$\Rightarrow \frac{2.0 \text{ eV}}{n - \frac{1}{2}} = \frac{2.8 \text{ eV}}{n + \frac{1}{2}} \Rightarrow 2.0n + 1.0 = 2.8n - 1.4 \Rightarrow n = 3$$

Thus, $E_3 = \left(3 - \frac{1}{2}\right)\hbar\omega = 2.5\hbar\omega = 2.0$ eV. Using Equation 40.43 for the angular frequency,

$$2.5\hbar\sqrt{\frac{k}{m}} = 2.0 \text{ eV} \Rightarrow k = m\left(\frac{2.0 \text{ eV}}{2.5\hbar}\right)^2 = (9.11\times10^{-31} \text{ kg})\left[\frac{(2.0 \text{ eV})(1.6\times10^{-19} \text{ J/eV})}{(2.5)(1.05\times10^{-34} \text{ J s})}\right]^2 = 1.4 \text{ N/m}$$

Section 40.10 Quantum-Mechanical Tunneling

40.23. Solve: A function $\psi(x)$ is a solution to the Schrödinger equation if

$$\frac{d^2\psi}{dx^2} = -\frac{2m}{\hbar^2}[E - U(x)]\psi(x)$$

Let $\psi(x) = A\psi_1(x) + B\psi_2(x)$, where $\psi_1(x)$ and $\psi_2(x)$ are both known to be solutions of the Schrödinger equation. The second derivative of $\psi(x)$ is

$$\frac{d^2\psi}{dx^2} = \frac{d^2}{dx^2}(A\psi_1(x) + B\psi_2(x)) = A\frac{d^2\psi_1}{dx^2} + B\frac{d^2\psi_2}{dx^2}$$

Since $\psi_1(x)$ and $\psi_2(x)$ are solutions, it must be the case that

$$\frac{d^2\psi_1}{dx^2} = -\frac{2m}{\hbar^2}[E - U(x)]\psi_1(x) \text{ and } \frac{d^2\psi_2}{dx^2} = -\frac{2m}{\hbar^2}[E - U(x)]\psi_2(x)$$

Using these results, the second derivative of $\psi(x)$ becomes

$$\frac{d^2\psi}{dx^2} = A\frac{d^2\psi_1}{dx^2} + B\frac{d^2\psi_2}{dx^2} = A\left(-\frac{2m}{\hbar^2}[E - U(x)]\psi_1(x)\right) + B\left(-\frac{2m}{\hbar^2}[E - U(x)]\psi_2(x)\right)$$

$$= -\frac{2m}{\hbar^2}[E - U(x)](A\psi_1(x) + B\psi_2(x))$$

$$= -\frac{2m}{\hbar^2}[E - U(x)]\psi(x)$$

Thus $\psi(x)$ is a solution to the Schrödinger equation.

40.27. Solve: From Equation 40.20, the wave functions for a particle in a box of length L are $\psi_n = A_n \sin(n\pi x/L)$. The wave function is nonzero only for $0 \le x < L$. The normalization requirement is

$$\int_{-\infty}^{\infty} |\psi(x)|^2 \, dx = A_n^2 \int_0^L \sin^2\left(\frac{n\pi x}{L}\right) dx = 1$$

Change the variable to $u = n\pi x/L$. Then, $dx = (L/n\pi)du$. The integration limits become $u = 0$ at $x = 0$ and $u = n\pi$ at $x = L$. The normalization integral, with the use of $\sin^2 u = \left(\frac{1}{2}\right)(1 - \cos 2u)$, becomes

$$1 = A_n^2 \frac{L}{n\pi} \int_0^{n\pi} \sin^2 u \, du = A_n^2 \frac{L}{n\pi} \int_0^{n\pi} \tfrac{1}{2}(1 - \cos 2u) du = A_n^2 \frac{L}{n\pi}\left[\tfrac{1}{2}u - \tfrac{1}{4}\sin 2u\right]_0^{n\pi} = \frac{LA_n^2}{2} \Rightarrow A_n = \sqrt{\frac{2}{L}}$$

40.29. Model: Model the particle as being confined in a rigid one-dimensional box of length L.
Visualize:

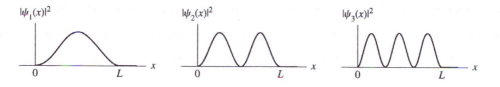

Solve: (a) The probability density is $|\psi_n(x)|^2 = (2/L)\sin^2(n\pi x/L)$. Graphs of $|\psi_1(x)|^2$, $|\psi_2(x)|^2$, and $|\psi_2(x)|^2$ are shown above.

(b) The particle is most likely to be found at x where $|\psi(x)|^2$ is a maximum. See table in part (d).

(c) The particle is least likely to be found at x where $|\psi(x)|^2 = 0$. See table in part (d).

(d) The probability of finding the particle in the left one-third of the box is the area under the $|\psi(x)|^2$ curve between $x = 0$ and $x = \frac{1}{3}L$. From examining the graphs, we can determine whether this is more than, less than, or equal to one-third of the total area. The results are shown in the table below.

n	Most likely	Least likely	Probability in left one-third
1	$\frac{1}{2}L$	0 and L	$<\frac{1}{3}$
2	$\frac{1}{4}L$ and $\frac{3}{4}L$	0, $\frac{1}{2}L$, and L	$>\frac{1}{3}$
3	$\frac{1}{6}L$, $\frac{3}{6}L$, and $\frac{5}{6}L$	0, $\frac{1}{3}L$, $\frac{2}{3}L$, and L	$=\frac{1}{3}$

(e) The probability of finding the particle in the range $0 \le x \le \frac{1}{3}L$ is

$$\text{Prob}(0 \le x \le \tfrac{1}{3}L) = \int_0^{L/3}|\psi_n(x)|^2\,dx = \frac{2}{L}\int_0^{L/3}\sin^2\left(\frac{n\pi x}{L}\right)dx$$

Change the variable to $u = n\pi x/L$. Then, $dx = (L/n\pi)du$. The integration limits become $u = 0$ at $x = 0$ m and $u = n\pi/3$ at $x = \frac{1}{3}L$. Then,

$$\text{Prob}(0 \le x \le \tfrac{1}{3}L) = \frac{2}{n\pi}\int_0^{n\pi/3}\sin^2 u\,du = \frac{2}{n\pi}\left[\tfrac{1}{2}u - \tfrac{1}{4}\sin 2u\right]_0^{n\pi/3} = \frac{1}{3} - \frac{1}{2n\pi}\sin\left(\frac{2n\pi}{3}\right)$$

The probability is 0.195 for $n = 1$, 0.402 for $n = 2$, and 0.333 for $n = 3$.

Assess: The results agree with the earlier estimates of the probability.

40.33. Solve: From Equations 40.45 and 40.46, $\psi_1(x) = A_1 e^{-x^2/2b^2}$, with $b^2 = \hbar/m\omega$ and $\omega = \sqrt{k/m}$. This is a solution to the Schrödinger equation *if* it's true that

$$\frac{d^2\psi_1}{dx^2} = -\frac{2m}{\hbar^2}\left[E - \frac{1}{2}kx^2\right]\psi_1(x)$$

where we've used $U(x) = \frac{1}{2}kx^2$. This equality is a hypothesis to be tested. To do so, we need the second derivative of $\psi_1(x)$:

$$\frac{d\psi_1(x)}{dx} = -\frac{A_1}{b^2}xe^{-x^2/2b^2} \Rightarrow \frac{d^2\psi_1(x)}{dx^2} = -\frac{A_1}{b^2}e^{-x^2/2b^2} + \frac{A_1}{b^4}x^2 e^{-x^2/2b^2} = -\left(\frac{1}{b^2} - \frac{x^2}{b^4}\right)\psi_1(x)$$

In the last step we used the definition of $\psi_1(x)$. Using the definition of b^2 and ω, this equation is

$$\frac{d^2\psi_1(x)}{dx^2} = -\left(\frac{m\omega}{\hbar} - \frac{m^2\omega^2 x^2}{\hbar^2}\right)\psi_1(x) = -\left(\frac{m\omega}{\hbar} - \frac{mkx^2}{\hbar^2}\right)\psi_1(x) = -\frac{2m}{\hbar^2}\left(\tfrac{1}{2}\hbar\omega - \tfrac{1}{2}kx^2\right)\psi_1(x)$$

Does this result for $d^2\psi_1(x)/dx^2$ equal $-(2m/\hbar^2)\left(E - \frac{1}{2}kx^2\right)\psi_1(x)$, as required by the Schrödinger equation? We can see by inspection that they are equal if $E = \frac{1}{2}\hbar\omega$. But this is the correct ground-state energy for the quantum harmonic oscillator. So we've shown that $\psi_1(x)$ *is* a solution of the Schrödinger equation.

40.35. Solve: (a) The ground-state wave function of the quantum harmonic oscillator is $\psi_1(x) = A_1 e^{-x^2/2b^2}$. Normalization requires

$$\int_{-\infty}^{\infty}|\psi_1(x)|^2\,dx = A_1^2\int_{-\infty}^{\infty}e^{-x^2/2b^2}\,dx = 1$$

Change the variable to $u = x/b$. Then, $dx = b\,du$. The integration limits don't change, so

$$1 = bA_1^2\int_{-\infty}^{\infty}e^{-u^2}\,du$$

The definite integral can be looked up in a table of integrals. The result is $\sqrt{\pi}$. Hence,

$$1 = bA_1^2\sqrt{\pi} = A_1^2\sqrt{\pi b^2} \Rightarrow A_1 = \frac{1}{(\pi b^2)^{1/4}}$$

(b) The forbidden region is both $x < -b$ and $x > b$. $|\psi_1(x)|^2$ is symmetrical about $x = 0$ m, so

$$\text{Prob}(x < -b \text{ or } x > b) = (2)\text{Prob}(x > b) = 2\int_b^\infty |\psi_1(x)|^2\, dx = \frac{2}{\sqrt{\pi b^2}}\int_b^\infty e^{-x^2/b^2}\, dx$$

(c) The integral of part (b) cannot be evaluated in closed form, but the answer can be found with a numerical integration. First, change the variable to $u = x/b$, making $dx = b\,du$. But unlike the variable change in part (c), this *does* change the lower limit of integration. Thus,

$$\text{Prob}(x < -b \text{ or } x > b) = \frac{2}{\sqrt{\pi b^2}}b\int_1^\infty e^{-u^2}\, du = \frac{2}{\sqrt{\pi}}\int_1^\infty e^{-u^2}\, du$$

The definite integral can be evaluated numerically with a calculator or computer, giving

$$\int_1^\infty e^{-u^2}\, du = 0.139$$

The probability of finding the harmonic oscillator in the forbidden region is $2(\pi)^{-\frac{1}{2}}(0.139) = 0.157 = 15.7\%$.

40.37. Model: The collisions with the ground are perfectly elastic.
Solve: (a) The classical probability density at position y of finding a ball that bounces between the ground and height h is given by Equation 40.32:

$$P_{\text{class}}(y) = \frac{2}{Tv(y)}$$

where $v(y)$ is the ball's velocity as a function of y and T is the period of oscillation. For a freely falling object, energy conservation gives

$$mgh = \tfrac{1}{2}mv^2 + mgy \Rightarrow v(y) = \sqrt{2g(h-y)}$$

The time $t = \frac{1}{2}T$ to reach a height h after a collision with the ground can be found from kinematics:

$$\Delta y = h = \tfrac{1}{2}gt^2 \Rightarrow t = \sqrt{\frac{2h}{g}}$$

$$\Rightarrow P_{\text{class}}(y) = \frac{2}{2\sqrt{2h/g}}\frac{1}{\sqrt{2g(h-y)}} = \sqrt{\frac{g}{2h}}\frac{1}{\sqrt{2g(h-y)}} = \frac{1}{2\sqrt{h}\sqrt{h-y}} = \left(\frac{1}{2h}\right)\frac{1}{\sqrt{1-(y/h)}}$$

(b)

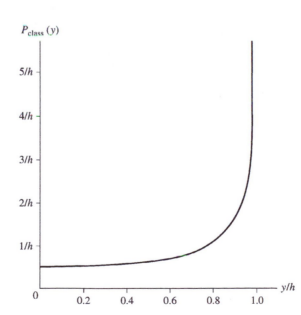

(c) The ball is most likely to be found near the upper turning point at $y = h$. This is because $v \rightarrow 0$ m/s at $y = h$ so the ball spends more time at this point. For the same reason, the ball spends the least time near the ground, where it is moving fastest, and thus the probability density is the least at $y = 0$ m.

40.39. Solve: Electrons are bound inside metals by an amount of energy called the work function E_0. This is the energy that must be supplied to lift an electron out of the metal. In this case, the expression $U_0 - E$ that appears in Equation 40.41 is E_0. Thus, the penetration distance is

$$\eta = \frac{\hbar}{\sqrt{2m(U_0 - E)}} = \frac{1.05 \times 10^{-34} \text{ J s}}{\sqrt{2(9.11 \times 10^{-31} \text{ kg})(4.3 \text{ eV})(1.6 \times 10^{-19} \text{ J/eV})}} = 9.38 \times 10^{-11} \text{ m}$$

From Equation 40.53, the probability that an electron will tunnel between the two aluminum pieces is

$$P_{\text{tunnel}} = e^{-2w/\eta} = e^{-2(50 \times 10^{-9} \text{ m})/(9.38 \times 10^{-11} \text{ m})} = e^{-1066} = 10^{-463} \approx 0$$

ATOMIC PHYSICS

Exercises and Problems

Section 41.1–41.2 The Hydrogen Atom

41.3. Solve: (a) The orbital angular momentum is $L = \sqrt{l(l+1)}\hbar$. Thus,

$$l(l+1) = \left(\frac{L}{\hbar}\right)^2 = \left(\frac{3.65 \times 10^{-34} \text{ J s}}{1.05 \times 10^{-34} \text{ J s}}\right)^2 = 12 \quad \Rightarrow \quad l = 3$$

This is an f electron (see Table 41.1).
(b) The l quantum number is required to be less than n. Thus, the minimum possible value of n for an electron in the f state is $n_{\text{min}} = 4$. From Equation 41.2, the corresponding minimum possible energy is

$$E_{\text{min}} = E_4 = -\frac{13.60 \text{ eV}}{4^2} = -0.850 \text{ eV}$$

41.5. Solve: A $6f$ state for a hydrogen atom corresponds to $n = 6$ and $l = 3$. Using Equation 41.2,

$$E_6 = \frac{-13.60 \text{ eV}}{6^2} = -0.378 \text{ eV}$$

The magnitude of the angular momentum is $L = \sqrt{l(l+1)}\hbar = \sqrt{3(3+1)}\hbar = \sqrt{12}\hbar$.

Section 41.3 The Electron's Spin

41.7. Solve: (a) A lithium atom has three electrons, two are in the $1s$ shell and one is in the $2s$ shell. The electron in the $2s$ shell has the following quantum numbers: $n = 2, l = 0, m = 0$, and m_s. m_s could be either $+\frac{1}{2}$ or $-\frac{1}{2}$. Thus, lithium atoms should behave like hydrogen atoms because lithium atoms could exist in the following two states: $\left(2, 0, 0, +\frac{1}{2}\right)$ and $\left(2, 0, 0, -\frac{1}{2}\right)$. Thus we would expect two lines.

(b) For a beryllium atom, we have two electrons in the $1s$ shell and two electrons in the $2s$ shell. The electrons in both the $1s$ and $2s$ states are filled. Because the two electron magnetic moments point in opposite directions, beryllium has *no* net magnetic moment and is not deflected in a Stern-Gerlach experiment. Thus we would expect only one line.

Section 41.4 Multielectron Atoms

Section 41.5 The Periodic Table of the Elements

41.9. Solve: Al, Ga, and In are all in the same column of the periodic table as boron. The electron configuration of Al $(Z=13)$ is $1s^2 2s^2 2p^6 3s^2 3p$. The electron configuration of Ga $(Z=31)$ is

$$1s^2 2s^2 2p^6 3s^2 3p^6 4s^2 3d^{10} 4p$$

The electron configuration of In $(Z=49)$ is

$$1s^2 2s^2 2p^6 3s^2 3p^6 4s^2 3d^{10} 4p^6 5s^2 4d^{10} 5p$$

Assess: Recall that the $4s$ subshell fills before the $3d$ subshell. Also note that the occupancy numbers (i.e., the superscripts) sum to give the total number of electrons for each atom. The superscript 1 is not written.

Section 41.6 Excited States and Spectra

41.13. Visualize: Please refer to Figure 41.15.
Solve: The diagram shows the energy levels for electrons in a multielectron atom. A lithium atom $(Z=3)$ in the ground state has two electrons in the $1s$ shell and 1 electron in the $2s$ shell. In other words, the ground-state electron configuration of lithium is $1s^2 2s$. The electron configuration for the first excited state is $1s^2 2p$ and for the second excited state is $1s^2 3s$.

Section 41.7 Lifetimes of Excited States

41.17. Solve: (a) If there are initially N_0 atoms in a state with lifetime τ, then the number remaining at time t is $N_{exc} = N_0 e^{-t/\tau}$. From Table 41.3, the $3p$ state has a lifetime $\tau = 17$ ns. Thus,

$$N_{exc} = (1.0 \times 10^6) e^{-10/17} = 555,000 \approx 5.6 \times 10^5$$

(b) Likewise at $t = 30$ ns, $N_{exc} = (1.0 \times 10^6) e^{-30/17} = 171,000 \approx 1.7 \times 10^5$.

(c) For $t = 100$ ns, $N_{exc} = (1.0 \times 10^6) e^{-100/17} = 3000 = 3.0 \times 10^3$.

41.19. Solve: (a) The number of atoms that have undergone a quantum jump to the ground state is

$$0.90 N_0 = 0.90 (1.0 \times 10^6) = 9.0 \times 10^5$$

Because each transition emits a photon, the number of emitted photons is also 9.0×10^5.
(b) 10% of the atoms remain excited at $t = 20$ ns. Thus, $N_{exc} = 0.10 N_0$, so

$$N_{exc} = N_0 e^{-t/\tau} \Rightarrow 0.10\, N_0 = N_0 e^{-20\,\text{ns}/\tau} \Rightarrow \ln(0.10) = \frac{-20\,\text{ns}}{\tau} \Rightarrow \tau = 8.7\,\text{ns}$$

Section 41.8 Stimulated Emission and Lasers

41.21. Solve: Since $1.0\,\text{mW} = 0.0010\,\text{W}$, the laser emits $E_{light} = 0.0010\,\text{J}$ of light energy per second. This energy consists of N photons. The energy of each photon is

$$E_{photon} = hf = \frac{hc}{\lambda} = 3.14 \times 10^{-19}\,\text{J}$$

Because $E_{light} = N E_{photon}$, the number of photons is

$$N = \frac{E_{light}}{E_{photon}} = \frac{0.0010\,\text{J}}{3.13 \times 10^{-19}\,\text{J}} = 3.2 \times 10^{15}$$

So, photons are emitted at the rate $3.2 \times 10^{15}\,\text{s}^{-1}$.

41.25. Solve: (a) For $s = 1$, $S = \sqrt{s(s+1)}\hbar = \sqrt{2}\hbar = 1.48 \times 10^{-34}$ J s.

(b) The spin quantum number is ms $m_s = -1, 0,$ or 1.

(c) The figure below shows the three possible orientations of $\vec{S}$.

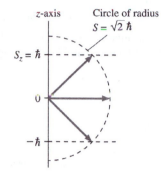

41.27. Solve: (a) Since $L_x^2 + L_y^2 + L_z^2 = L^2$, $L_x^2 + L_y^2 = L^2 - L_z^2$ and $\sqrt{L_x^2 + L_y^2} = \sqrt{L^2 - L_z^2}$. For a hydrogen atom with $l = 2$, the magnitude of L^2 is always $l(l+1)\hbar^2$ or $6\hbar^2$. The value of L_z is $m\hbar$, where m is an integer between $-l$ and l. The minimum of $\sqrt{L^2 - L_z^2}$ occurs when L_z^2 is maximum, which occurs for $m = \pm l$. Thus, the minimum value of $\sqrt{L^2 - L_z^2}$

$$\sqrt{L^2 - L_z^2}\Big|_{\min} = \sqrt{l(l+1)\hbar^2 - (\pm l\hbar)^2} = \sqrt{6\hbar^2 - (\pm 2\hbar)^2} = \sqrt{2}\hbar$$

(b) The minimum value of L_z occurs when $m = 0$, so the maximum value of $\sqrt{L_x^2 + L_y^2}$ is

$$\sqrt{L^2 - L_z^2}\Big|_{\max} = \sqrt{6\hbar^2 - (0\hbar)^2} = \sqrt{6}\hbar$$

41.33. Solve: (a) From Equation 41.7, the $2p$ radial wave function is

$$R_{2p}(r) = \frac{A_{2p}}{2a_B} r e^{-r/2a_B}$$

The graph of $R_{2p}(r)$ is seen to have a single maximum.

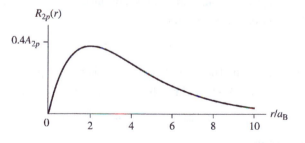

(b) $R_{2p}(r)$ is a maximum at the point where $dR_{2p}/dr = 0$. The derivative is

$$\frac{dR_{2p}}{dr} = \frac{A_{2p}}{2a_B}\left[e^{-r/2a_B} - \frac{r}{2a_B}e^{-r/2a_B} \right] = \frac{A_{2p}}{2a_B}\left[1 - \frac{r}{2a_B} \right]e^{-r/2a_B}$$

The derivative is zero at $r = 2a_B$, so $R_{2p}(r)$ is a maximum at $r = 2a_B$.

(c) The absolute square of the radial wave function $\left|R_{2p}\right|^2$ gives the probability *per unit volume* of finding the electron at a given distance from the origin (see figure below). We can imagine traveling away from the origin along a radial ray and inspecting differential volume elements $\delta V = \delta r^3$ as we go to determine the probability of finding the electron in the differential boxes as a function of r.

The probability density $P_{2p}(r)$ gives the probability *per unit length* of finding the electron in a shell of thickness δr and radius r (see figure below). Again, we can imagine traveling away from the origin along a ray, but this time inspecting a spherical shell of thickness δr to find the probability of finding the electron within this shell.

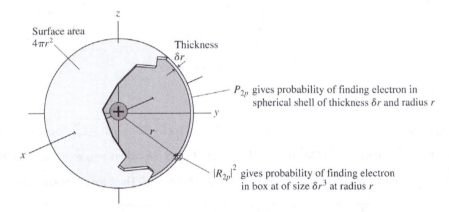

Thus, the two quantities differ by a factor of the surface area of the spherical shell:

$$P_{2p}(r) = 4\pi r^2 \left|R_{2p}(r)\right|^2$$

The r^2 factor for the surface area expresses the fact that there is more volume in which to find the electron in a spherical shell of thickness δr than in a box of dimensions δr^3. This factor pushes the maximum of $P_{2p}(r)$ out farther from the origin, which is why $P_{2p}(r)$ peaks at $r = 4a_B$ rather than at $r = 2a_B$.

41.35. Solve: The electron configuration in the ground state of K ($Z = 19$) is

$$1s^2 2s^2 2p^6 3s^2 3p^6 4s$$

This means that all the states except the $4s$ state are completely filled. For Sc ($Z = 21$), all the states up to the $4s$ state are completely filled. The electron configuration is

$$1s^2 2s^2 2p^6 3s^2 3p^6 4s^2 3d$$

The d subshell has only a single electron. The case of Cu ($Z = 29$) is an exception, because it prefers a completely filled $3d$ subshell with 10 electrons and a $4s$ state with only a single electron. The electron configuration is therefore

$$1s^2 2s^2 2p^6 3s^2 3p^6 4s3d^{10}$$

The ground-state electron configurations of Ge ($Z = 32$) is

$$1s^2 2s^2 2p^6 3s^2 3p^6 4s^2 3d^{10} 4p^2$$

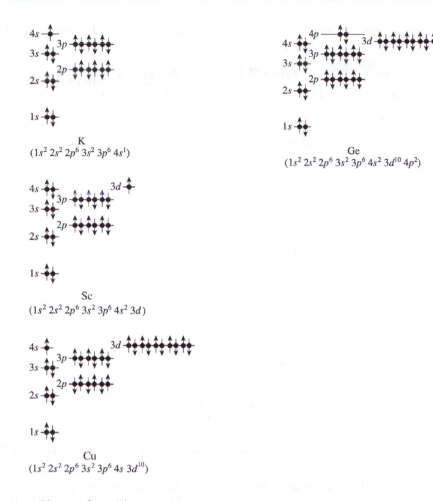

K
$(1s^2\, 2s^2\, 2p^6\, 3s^2\, 3p^6\, 4s^1)$

Ge
$(1s^2\, 2s^2\, 2p^6\, 3s^2\, 3p^6\, 4s^2\, 3d^{10}\, 4p^2)$

Sc
$(1s^2\, 2s^2\, 2p^6\, 3s^2\, 3p^6\, 4s^2\, 3d\,)$

Cu
$(1s^2\, 2s^2\, 2p^6\, 3s^2\, 3p^6\, 4s\, 3d^{10})$

41.37. Visualize: Please refer to Figure 41.24.
Solve: (a) The allowed transitions are those with $\Delta l = \pm 1$. Starting from the $6s$ state, the only allowed transitions are to p states that are lower in energy and that are not already filled (i.e., the $2p$ states). Referring to Figure 41.24, we see that these states are the $5p$, $4p$, and $3p$ states.
(b) Once the states are identified, the wavelength is $\lambda = hc/\Delta E = 1240$ eV nm$/\Delta E$.

Transition	ΔE	λ
$6s \rightarrow 5p$	0.17 eV	7.29 μm
$6s \rightarrow 4p$	0.76 eV	1.63 μm
$6s \rightarrow 3p$	2.41 eV	515 nm

41.39. Visualize: Please refer to Figure 41.24.
Solve: A wavelength of 818 nm corresponds to an energy of

$$E = \frac{hc}{\lambda} = \frac{1240 \text{ eV nm}}{818 \text{ nm}} = 1.52 \text{ eV}$$

From Figure 41.24, the transition that obeys the selection rule $\Delta l = 1$ and has a magnitude around 1.5 eV is $3p \rightarrow 3d$. Note that $E_{3d} - E_{3p} = 3.620$ eV $- 2.104$ eV $= 1.516$ eV, which, within the precision of the measurement of the wavelength, is the same as the energy of the emitted photon. The atom was therefore excited to the $3d$ state from the ground state. Thus the minimum kinetic energy of the electron was

$$\frac{1}{2}mv^2 \geq 3.62 \text{ eV} \quad \Rightarrow \quad v \geq \sqrt{\frac{2(3.62 \text{ eV})(1.602 \times 10^{-19} \text{ J/eV})}{9.11 \times 10^{-31} \text{ kg}}} = 1.13 \times 10^6 \text{ m/s}$$

41.43. Model: We have a one-dimensional rigid box with infinite potential walls and a length L of 0.50 nm.

Solve: **(a)** From Equation 40.22, the lowest energy level is

$$E_1 = \frac{h^2}{8mL^2} = \left(\frac{(6.63 \times 10^{-34} \text{ J s})^2}{8(9.11 \times 10^{-31} \text{ kg})(5.0 \times 10^{-10} \text{ m})^2} \right)\left(\frac{1.00 \text{ eV}}{1.60 \times 10^{-19} \text{ J}} \right) = 1.51 \text{ V}$$

The next two levels are $E_2 = (2^2)E_1 = 6.04$ eV and $E_3 = (3^2)E_1 = 13.6$ eV. The Pauli principle allows only two electrons in each of these energy levels; one with spin up and one with spin down, so five electrons fill the $n=1$ and $n=2$ levels, with the fifth electron going into the $n=3$ level. The energy-level diagram is shown below:

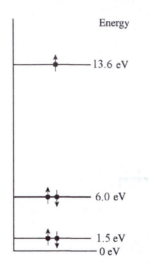

(b) The ground-state energy of these five electrons is $E = 2E_1 + 2E_2 + E_3 = 2(1.51 \text{ eV}) + 2(6.40 \text{ eV}) + 13.6 \text{ eV} = 28.7 \text{ eV}$

41.45. Solve: **(a)** From Table 41.3, the lifetime of the $2p$ state of hydrogen is $\tau = 1.6$ ns. The decay rate is

$$r = \frac{1}{\tau} = \frac{1}{1.6 \times 10^{-9} \text{ s}} = 6.25 \times 10^8 \text{ s}^{-1} \approx 6.3 \times 10^8 \text{ s}^{-1}$$

(b) From Equation 41.25, the number of excited atoms left at time t is $N_{exc} = N_0 e^{-t/\tau}$. If 10% of a sample decays, then 90% of the atoms in the sample are still excited. That is, $N_{exc} = 0.90 N_0$. The time for this to occur is:

$$N_{exc} = 0.90 N_0 = N_0 e^{-t/\tau} \quad \Rightarrow \quad e^{-t/\tau} = 0.90 \quad \Rightarrow \quad t = -\tau \ln(0.90) = 0.17 \text{ ns}$$

41.49. Solve: If there are N_{exc} atoms in the excited state, the *rate* of decay is rN_{exc}, where $r = 1/\tau$ is the decay rate. Each decay generates one photon, so the rate of photon emission (photons per second) is the same as the rate of decay (decays per second). Hence, the number of photons emitted per second is

$$rN_{exc} = \frac{N_{exc}}{\tau} = \frac{1.0 \times 10^9}{20 \times 10^{-9} \text{ s}} = 5.0 \times 10^{16} \text{ s}^{-1}$$

Thus, in one second, there are 5.0×10^{16} photons emitted.

NUCLEAR PHYSICS

Exercises and Problems

Section 42.1 Nuclear Structure

42.1. Model: The nucleus is composed of Z protons and $A - Z$ neutrons.

Solve: (a) ^{3}He has $Z = 2$ protons and $3 - 2 = 1$ neutron.

(b) ^{32}P has $Z = 15$ protons and $32 - 15 = 17$ neutrons.

(c) ^{32}S has $Z = 16$ protons and $32 - 16 = 16$ neutrons.

(d) ^{238}U has $Z = 92$ protons and $238 - 92 = 146$ neutrons.

42.5. Visualize: The masses of the nuclei are found by subtracting the mass of Z electrons from the atomic masses in Appendix C, and then converting to kg.

$$m(^6\text{Li}) = 6.015121 \text{ u} - 3(0.000548) \text{ u} = 6.0135 \text{ u}$$

$$m(^{207}\text{Pb}) = 206.975871 \text{ u} - 82(0.000548) \text{ u} = 206.9309 \text{ u}$$

The radii are computed from $r = r_0 A^{1/3}$ where $r_0 = 1.2$ fm. The densities are computed from

$$\rho = \frac{m}{\frac{4}{3}\pi r^3}$$

Solve: (a) For ^{6}Li:

$$m = 6.0135 \text{ u} \left(\frac{1.661 \times 10^{-27} \text{ kg}}{1 \text{ u}} \right) = 9.988 \times 10^{-27} \text{ kg}$$

$$r = r_0 A^{1/3} = (1.2 \times 10^{-15} \text{ m})(6)^{1/3} = 2.18 \times 10^{-15} \text{ m} \approx 2.2 \times 10^{-15} \text{ m}$$

$$\rho = \frac{m}{\frac{4}{3}\pi r^3} = \frac{9.988 \times 10^{-27} \text{ kg}}{\frac{4}{3}\pi (2.18 \times 10^{-15} \text{ m})^3} = 2.3 \times 10^{17} \text{ kg/m}^3$$

(b) For ^{207}Pb:

$$m = 206.9309 \text{ u} \left(\frac{1.661 \times 10^{-27} \text{ kg}}{1 \text{ u}} \right) = 3.437 \times 10^{-25} \text{ kg}$$

$$r = r_0 A^{1/3} = (1.2 \times 10^{-15} \text{ m})(207)^{1/3} = 7.10 \times 10^{-15} \text{ m} \approx 7.1 \times 10^{-15} \text{ m}$$

$$\rho = \frac{m}{\frac{4}{3}\pi r^3} = \frac{3.437 \times 10^{-25} \text{ kg}}{\frac{4}{3}\pi (7.10 \times 10^{-15} \text{ m})^3} = 2.3 \times 10^{17} \text{ kg/m}^3$$

Assess: The densities are very similar because the nucleons are tightly packed in both cases. All nuclei have similar densities. This is also a typical density for a neutron star.

Section 42.2 Nuclear Stability

42.7. Solve: (a) $A = 36$, for which ^{36}S and ^{36}Ar are stable.

(b) Nuclei with $A = 8$ and $A = 5$ have no stable nuclei.

42.11. Solve: From Equation 42.6, the binding energy for ^{14}O is

$$B = Zm_H + Nm_n - m_{atom} = 8(1.00783 \text{ u}) + 6(1.00866 \text{ u}) - 14.008595 \text{ u}$$
$$= 0.106005 \text{ u} \times 931.49 \text{ MeV/u} = 98.7426 \text{ MeV}$$

The binding energy per nucleon is $\frac{1}{14}(98.7426 \text{ MeV}) = 7.05 \text{ MeV}$.

For ^{16}O, the binding energy is

$$B = Zm_H + Nm_n - m_{atom} = 8(1.00783 \text{ u}) + 8(1.00866 \text{ u}) - 15.994915 \text{ u}$$
$$= 0.137005 \text{ u} \times 931.49 \text{ MeV/u} = 127.6188 \text{ MeV}$$

The binding energy per nucleon is $\frac{1}{16}(127.6188 \text{ MeV}) = 7.98 \text{ MeV}$. Thus, ^{16}O is slightly more tightly bound.

Section 42.3 The Strong Force

42.15. Solve: From Figure 42.8, the nuclear potential energy at $r = 1.0$ fm is

$$U_{nuclear} = -50 \text{ MeV} = -(50 \text{ MeV})(1.6 \times 10^{-19} \text{ J/eV}) = -8.0 \times 10^{-12} \text{ J}$$

The gravitational potential energy is

$$U_{grav} = -\frac{Gm^2}{r} = -\frac{(6.67 \times 10^{-11} \text{ Nm}^2/\text{kg}^2)(1.67 \times 10^{-27} \text{ kg})^2}{1.0 \times 10^{-15} \text{ m}} = -1.86 \times 10^{-49} \text{ J}$$

$$\Rightarrow \frac{U_{grav}}{U_{nuclear}} = \frac{1.86 \times 10^{-49} \text{ J}}{8.0 \times 10^{-12} \text{ J}} = 2.3 \times 10^{-38}$$

Section 42.4 The Shell Model

42.17. Solve: (a) The $A = 14$ nuclei listed in Appendix C are ^{14}C, ^{14}N, and ^{14}O. ^{14}C has $Z = 6$, so $N = 8$. ^{14}N has $Z = 7$, $N = 7$, and ^{14}O has $Z = 8$, $N = 6$. These 3 nuclei are the $A = 14$ isobars.

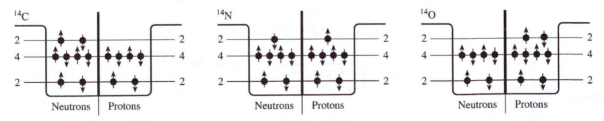

(b) ^{14}N is a stable nucleus, but ^{14}C and ^{14}O are radioactive. ^{14}C undergoes beta-minus decay and ^{14}O undergoes beta-plus decay. In beta-minus decay, a neutron within the nucleus changes into a proton and an electron. In beta-plus decay, a proton changes into a neutron and a positron.

Assess: The above decays for ^{14}C and ^{14}O are consistent with the fact that the line of stability follows the $N = Z$ line for $Z < 16$.

Section 42.5 Radiation and Radioactivity

42.19. Model: The number of radioactive atoms decreases exponentially with time.
Solve: The mass of remaining ^{131}Ba nuclei at time t is

$$N = N_0 \left(\frac{1}{2}\right)^{t/t_{1/2}}$$

(a) After 1 day, $N = (250\ \mu g)(1/2)^{1\ \text{day}/12\ \text{days}} = 236\ \mu g$.

(b) After 10 days, $N = (250\ \mu g)(1/2)^{10\ \text{days}/12\ \text{days}} = 140\ \mu g$.

(c) After 100 days, $N = (250\ \mu g)(1/2)^{100\ \text{days}/12\ \text{days}} = 0.775\ \mu g$.

Assess: Do not think that if half the nuclei decay during one half-life, then all will decay in two half-lives.

42.21. Model: The number of atoms decays exponentially with time.
Solve: The number of remaining radioactive atoms at $t = 50$ min and $t = 200$ min is

$$N_{50} = N_0 \left(\frac{1}{2}\right)^{t/t_{1/2}} = (1.0 \times 10^{10}) \left(\frac{1}{2}\right)^{(50\ \text{min})/(100\ \text{min})} = 7.07 \times 10^9\ \text{atoms}$$

$$N_{200} = N_0 \left(\frac{1}{2}\right)^{t/t_{1/2}} = (1.0 \times 10^{10}) \left(\frac{1}{2}\right)^{(200\ \text{min})/(100\ \text{min})} = 2.50 \times 10^9\ \text{atoms}$$

Thus the number that decay between 50 min and 200 min is

$$N_{\text{decay}} = N_{50} - N_{200} = 4.6 \times 10^9\ \text{atoms}$$

Each decay emits an alpha particle, so there are 4.6×10^9 alphas emitted.

Section 42.6 Nuclear Decay Mechanisms

42.25. Solve: (a) The decay is $X \rightarrow {}^{224}\text{Ra} + {}^{4}\text{He}$, so $X = {}^{228}\text{Th}$.

(b) The decay is $X \rightarrow {}^{207}\text{Pb} + e^- + \overline{v}$, so $X = {}^{207}\text{Tl}$.

(c) The decay is $^{7}\text{Be} + e^- \rightarrow X + v$, so $X = {}^{7}\text{Li}$.

(d) The decay is $X \rightarrow {}^{60}\text{Ni} + \gamma$, so $X = {}^{60}\text{Ni}$.

42.29. Model: Assume the energy released goes into the alpha particle's kinetic energy so that $K_\alpha = 5.52$ MeV.
Visualize: Use Equation 42.15: $K_\alpha = (m_X - m_Y - m_{\text{He}})c^2 = 5.52$ MeV. We are also given $m_X + m_Y = 452$ u.
Solve:

$$(m_X - m_Y - m_{\text{He}})c^2 = 5.52\ \text{MeV}$$

$$(m_X - m_Y - m_{\text{He}}) = 5.52\ \text{MeV}/c^2 \left(\frac{1\ \text{u}}{931.49\ \text{MeV}/c^2}\right) = 0.00593\ \text{u}$$

$$m_X - m_Y = 0.00593\ \text{u} + m_{\text{He}} = 0.00593\ \text{u} + 4.00260\ \text{u} = 4.00853\ \text{u}$$

We now have a system of two equations in two unknowns which we will add together to solve for m_X.

$$\begin{cases} m_X - m_Y = 4.00853\ \text{u} \\ m_X + m_Y = 452\ \text{u} \end{cases}$$

$$2m_X = 456\ \text{u}$$

$$m_X = 228\ \text{u}$$

Looking in Appendix C for a nucleus with a mass of 228 u which decays by α decay, we find ^{228}Th.

Assess: ^{228}Th is a step in the ^{232}Th decay series. ^{238}Th has a half-life of 1.9 yr and decays by α radiation releasing 5.52 MeV of energy.

42.31. Model: The decay is $^{24}\text{Na} \rightarrow {}^{24}\text{Mg} + e^- + \bar{\nu}$.

Solve: Beta-minus decay leaves the daughter atom as a positive ion. However, the mass of the ion plus the mass of the escaping electron are the mass of a neutral atom, which is what is tabulated in Appendix C. Thus the mass loss is the mass difference between the two neutral atoms. In Appendix C we find that $m(^{24}\text{Na}) = 23.990961$ u and $m(^{24}\text{Mg}) =$ 23.985042 u. The energy released in the beta-minus decay corresponds to a mass change of 0.005919 u. The energy released is

$$E = \Delta mc^2 = (0.005919 \text{ u})c^2 \times \frac{931.49 \text{ MeV}/c^2}{1 \text{ u}} = 5.51 \text{ MeV}$$

Assess: This energy is shared between the electron and the antineutrino.

Section 42.7 Biological Applications of Nuclear Physics

42.35. Model: The radiation dose in units of rems is a combination of deposited energy and biological effectiveness. The RBE for beta radiation is 1.5.

Solve: 1 Gy is defined as 1.00 J/kg of absorbed energy. In the case of the 50 kg worker, the energy absorbed per kg is

$$\frac{20 \times 10^{-3} \text{ J}}{50 \text{ kg}} = 4.0 \times 10^{-4} \text{ J/kg}$$

This corresponds to a dose of

$$4.0 \times 10^{-4} \text{ J/kg} \times \frac{1 \text{ Gy}}{1.00 \text{ J/kg}} = 4.0 \times 10^{-4} \text{ Gy}$$

The dose equivalent in rems is $(4.0 \times 10^{-4} \text{ Gy})(1.5) = 0.060 \text{ Sv} = 60 \text{ mrem}$.

42.37. Model: Assume the ^{197}Au nucleus remains at rest.

Visualize: The radii of the alpha particle and the gold nucleus are

$$r_\alpha = (1.2 \text{ fm})(4)^{1/3} = 1.90 \text{ fm} \qquad r_{\text{Au}} = (1.2 \text{ fm})(197)^{1/3} = 6.98 \text{ fm}$$

If the alpha just touches the surface of the gold nucleus, the distance between their centers is $r_f = 1.90 \text{ fm} + 6.98 \text{ fm} =$ 8.88 fm.

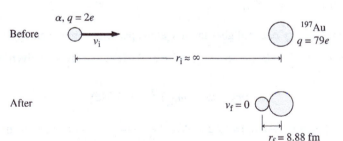

Solve: (a) The energy conservation equation $K_f + U_f = K_i + U_i$ is

$$0 \text{ J} + \frac{1}{4\pi\varepsilon_0} \frac{(2e)(79e)}{8.88 \times 10^{-15} \text{ fm}} = \frac{1}{2}mv_i^2 + 0 \text{ J}$$

$$\Rightarrow v_i = \sqrt{\frac{2(9.0 \times 10^9 \text{ N m}^2/\text{C}^2)(158)(1.60 \times 10^{-19} \text{ C})^2}{(4 \times 1.661 \times 10^{-27} \text{ kg})(8.88 \times 10^{-15} \text{ fm})}} = 3.5 \times 10^7 \text{ m/s}$$

(b) The energy of the alpha particle is

$$K = \frac{1}{2}mv_i^2 = \frac{1}{2}(4 \times 1.661 \times 10^{-27} \text{ kg})(3.5 \times 10^7 \text{ m/s})^2$$

$$= 4.09 \times 10^{-12} \text{ J} \times (1 \text{ MeV}/1.60 \times 10^{-13} \text{ J}) = 26 \text{ MeV}$$

42.39. Solve: (a) The sun's mass of 1.99×10^{30} kg is unchanged, but it assumes the density of nuclear matter, which we found to be $\rho_{nuc} = 2.3 \times 10^{17}$ kg/m^3. The volume of the collapsed sun is

$$V = \frac{4}{3}\pi r^3 = \frac{M_S}{\rho_{nuc}} = \frac{1.99 \times 10^{30} \text{ kg}}{2.3 \times 10^{17} \text{ kg/m}^3} = 8.65 \times 10^{12} \text{ m}^3$$

Thus its radius is

$$r = \left(\frac{3(8.65 \times 10^{12} \text{ m}^3)}{4\pi} \right)^{1/3} = 12{,}700 \text{ m} = 12.7 \text{ km}$$

(b) We can use the conservation of angular momentum to find the new rotational period:

$$(I\omega)_{after} = (I\omega)_{before} \Rightarrow \omega_{after} = \frac{I_{before}}{I_{after}} \omega_{before} = \frac{\frac{2}{5}M_S R_S^2}{\frac{2}{5}M_S r^2} \omega_{before}$$

$$\Rightarrow \frac{2\pi}{T_{after}} = \left(\frac{R_S}{r} \right)^2 \frac{2\pi}{T_{before}} \Rightarrow T_{after} = \left(\frac{1.27 \times 10^4 \text{ m}}{6.96 \times 10^8 \text{ m}} \right)^2 (27 \text{ days}) = 7.8 \times 10^{-4} \text{ s} = 780 \text{ } \mu s$$

42.43. Visualize: Please refer to Figure 42.6 for the graph of the binding energy versus mass number.
Solve: For a ^{4}He nucleus, the binding energy per nucleon is 7.2 MeV. Because $A = 4$, the binding energy of the ^{4}He nuclei is 7.2 MeV $\times$ 4 = 28.8 MeV. Three ^{4}He nuclei thus have a total energy of 28.8 MeV $\times$ 3 = 86.4 MeV. For the ^{12}C nucleus, the binding energy per nucleon is 7.7 MeV. Because $A = 12$, the binding energy of ^{12}C is 7.7 MeV $\times$ 12 = 92.4 MeV. When three ^{4}He nuclei fuse together to form a ^{12}C nucleus, the total energy released is 92.4 MeV − 86.4 MeV = 6.0 MeV.

42.45. Solve: The radius of a ^{238}U nucleus is

$$r = r_0 A^{1/3} = (1.2 \text{ fm})(238)^{1/3} = 7.436 \text{ fm}$$

For the de Broglie wavelength to be equal to the diameter of the ^{238}U nucleus, $p = h/\lambda = h/2r$. Hence,

$$K = \frac{p^2}{2m} = \left(\frac{h}{2r} \right)^2 \frac{1}{2m} = \left(\frac{6.63 \times 10^{-34} \text{ J s}}{2 \times 7.436 \times 10^{-15} \text{ m}} \right)^2 \frac{1}{2(4 \times 1.67 \times 10^{-27} \text{ kg})} \times \frac{1 \text{ eV}}{1.6 \times 10^{-19} \text{ J}} = 0.93 \text{ MeV}$$

42.49. Model: The decay is ^{223}Ra $\rightarrow$ ^{219}Rn + ^{4}He.
Solve: (a) The energy released in the above decay is

$$E = [m(^{223}\text{Ra}) - m(^{219}\text{Rn}) - m(^4\text{He})] \times 931.49 \text{ MeV/u}$$
$$= [223.018499 \text{ u} - 219.009477 \text{ u} - 4.002602 \text{ u}] \times 931.49 \text{ MeV/u} = 5.98 \text{ MeV}$$

That is, each α-particle is released with an energy of

$$5.98 \text{ MeV} = 5.98 \times 10^6 \text{ eV} \times 1.6 \times 10^{-19} \text{ J/eV} = 9.57 \times 10^{-13} \text{ J}$$

The amount of energy needed to raise the temperature of 100 mL of water at 18°C to 100°C is

$$Q = mc\Delta t = (0.10 \text{ L}) \left(\frac{1 \text{ kg}}{1 \text{ L}} \right) (4190 \text{ J/kg K})(100 \text{ K} - 18 \text{ K}) = 34{,}400 \text{ J}$$

The number of decays we need to generate this amount of energy is

$$\frac{34{,}400 \text{ J}}{9.57 \times 10^{-13} \text{ J}} = 3.59 \times 10^{16}$$

The total number of radium atoms in the cube at $t = 0$ s is

$$N_0 = \frac{1\,\text{g}}{223\,\text{g/mol}} \times 6.02 \times 10^{23}\ \text{atoms/mol} = 2.70 \times 10^{21}\ \text{atoms}$$

The number of needed decays is very small compared to N_0, so we can write

$$\frac{dN}{dt} \approx \frac{\Delta N}{\Delta t} = -rN \quad \Rightarrow \quad \Delta t = -\frac{\Delta N}{rN} = -\tau \frac{\Delta N}{N} = -\frac{t_{1/2}}{\ln 2} \frac{\Delta N}{N}$$

where we used $r = 1/\tau$. $\Delta N = -3.59 \times 10^{16}$, with the negative sign due to the fact that the number of radium atoms decreases. Since $t_{1/2} = 11.43\ \text{days} = 9.876 \times 10^5\ \text{s}$, the time for 3.59×10^{16} decays is

$$\Delta t = -\frac{9.876 \times 10^5\ \text{s}}{\ln 2} \frac{(-3.59 \times 10^{16})}{2.70 \times 10^{21}} = 18.9\ \text{s} \approx 19\ \text{s}$$

(b) It's possible that an alpha-particle collision will break the molecular bond in a *very* small number of H_2O molecules, causing H_2 gas and O_2 gas to bubble out of the water. The water that remains in the container has not changed or been altered.

42.53. Model: The number of ^{40}K atoms decays exponentially with time.
Solve: Suppose the number of ^{40}K atoms is N_0 at the time the lava solidifies. There are no ^{40}Ar atoms at that time. As time passes, 11% of the ^{40}K that decays goes into ^{40}Ar, but the total number of atoms locked inside the lava is unchanged. That is, $N_{Ar} = (0.11)(N_0, -N_K)$ where N_K is the remaining number of ^{40}K atoms. Dividing by N_K, we have

$$\frac{N_{Ar}}{N_K} = (0.11)\left(\frac{N_0}{N_K} - 1\right) \Rightarrow \frac{0.013}{0.11} + 1 = \frac{N_0}{N_K} \Rightarrow \frac{N_K}{N_0} = \frac{1}{1.118} = 0.894$$

That is, 89.4% of the ^{40}K remains at a time when the Ar/K ratio is 0.013. The ^{40}K decays as

$$N_K = N_0 \left(\frac{1}{2}\right)^{t/t_{1/2}} \Rightarrow \frac{t}{t_{1/2}} \ln(1/2) = \ln(N_K/N_0)$$

$$\Rightarrow t = (1.28\ \text{billion years}) \frac{\ln(0.894)}{\ln(0.50)} = 0.21\ \text{billion years} = 210\ \text{million years}$$

42.55. Solve: Since the dose equivalent in Sv is the dose in Gy times the RBE, the dose is

$$\frac{0.30 \times 10^{-3}\ \text{Sv}}{0.85} = 3.53 \times 10^{-4}\ \text{Gy} \times \frac{1.00\ \text{J/kg}}{1\ \text{Gy}} = 3.53 \times 10^{-4}\ \text{J/kg}$$

Since only one-fourth of the body received x-ray exposure, the total amount of energy received is

$$3.53 \times 10^{-4}\ \text{J/kg} \times \frac{1}{4} \times 60\ \text{kg} = 5.29 \times 10^{-3}\ \text{J}$$

The energy of each photon is

$$10\ \text{keV} \times \frac{1.6 \times 10^{-19}\ \text{J}}{1\ \text{eV}} = 1.6 \times 10^{-15}\ \text{J}$$

Thus, the number of x-ray photons absorbed by the body is

$$\frac{5.29 \times 10^{-3}\ \text{J}}{1.6 \times 10^{-15}\ \text{J}} = 3.3 \times 10^{12}$$